REDWATER

Aubrey Kerr

...We were threshing a mile east ... We saw this big plume of black smoke, and we knew that rig had struck oil ...

Walter Hrynchuk

ISBN 0-9698560-0-8

Published by
S. A. Kerr
912 - 80 Ave. S.W.
Calgary, Alberta, T2V 0V3

First Printing, 1994

Also by the author:
ATLANTIC No. 3, 1948 © 1986
CORRIDORS OF TIME © 1988
LEDUC © 1991

Design and Typesetting:
By Design Services
Calgary, Alberta T3A 1K2

Printed and bound in Canada by
Kromar Printing Limited
Winnipeg, Manitoba
Canada

TO

THOSE PIONEERS
WHO SAW THE BIRTH OF OIL
AND
THEIR DESCENDANTS
WHO WILL TILL THE SOIL LONG AFTER
OIL IS FORGOTTEN.

Table of Contents

	Preface	vii
1	Bird's Eye View	1
2	No Oil in Alberta, Eh?	7
3	Follow-ups	27
4	For the Geologist	59
5	For the Engineer	77
6	The Land: Freehold	127
7	The Land: Crown	155
8	The People	177
	Profiles	205
	Appendices	277
	Bibliography	373
	Acknowledgements	375
	Index	379

Preface

"...When I have fears that I may cease to be before my pen has gleaned my teeming brain..."

Keats' Sonnet No.7

The main reason for documenting "REDWATER" is embodied in the above quotation passed on to me by Doug Lavoie. Secondly, I have been encouraged by several friends, notably John Andrichuk: "This will be your best book yet." Thirdly, I sense an obligation to chronicle Redwater before too many players have left the stage. Fourthly, and most suasive: the spur of the hunt, chasing down stories - no matter how outrageous, digging up records and, in the process, getting acquainted with many wonderful people.

My research has resulted in a wealth of information - from the technical to the sociological - much of which has never appeared in print. REDWATER captures as much of the information as possible to ensure that it is not lost for all time. I have attempted to present this pot pourri of research, to my pot pourri of readers, in as manageable a way as possible.

Although I interwove as much of the character and background of the people as possible into the text, a wealth of interesting information still remained. In appreciation of the many individuals who have contributed their valuable time to my research, and for posterity's sake, I have taken the liberty of including this material in the form of Blue Pages following Chapter 8. Enjoy!

A key activist in this interweaving, mixing and matching of material has been my editor, Carol Weaver, who diligently, and forthrightly, kept the author cognizant of his readership without bruising his ego. Without her, the book would not have come into being. Also valued is the contribution of Sophia Rybkowski, who shepherded the book from its beginning, typing endless drafts while applying her unique brand of proof-reading. The excellent editing skills of Donald J. Glass were once again put to work. Don made some worthwhile changes, at the same time vigorously challenging the author's "comma culture".

The heartbeat of the project lies with its sponsors: The unfailing, generous contribution of office space and facilities by Encor Energy

Corporation and, most recently, Home Oil Company Limited; financial help from a private oil company, the Alberta Historical Resources Foundation and New Horizons.

Townspeople, such as Anne and Lester Key, have contributed greatly to all aspects of the narrative and truly represent their diverse community. Up to the time of his resignation, Keith Miller, Redwater's Town Manager, had taken a personal interest in the research.

Other organizations and individuals to be acknowledged are:

- Imperial Oil Limited for access to their files.
- Energy Resources Conservation Board staff in Calgary and Edmonton.
- By Design Services, Sandy Irvine, desktop publisher.
- Doreen McArthur's expert advice in layout and page-making.
- Evelyn De Mille for her staunch support over the years.
- Gordon Hawkins and Hubert Somerville for invaluable advice on land tenure, dispelling myths and contributing to a true account of events from as far back as 1867.
- Geologists: John Andrichuk, Jean McLaws, Don Hoyda, Doug Lavoie, and Doug Carsted.
- My final and special thanks are reserved for my wife, Elsie, my toughest critic. Her cautionary advice on how to avoid literary and financial pitfalls has proven to be indispensable.

As with previous utterances, I take complete responsibility for the text, knowing full well that the credibility of some folkloric accounts may be questionable.

1 BIRD'S EYE VIEW

...a wide spot in the road...

Coral reefs flourished some 400,000,000 years ago in warm saline, shallow seas. This may stretch one's credulity, but the evidence is buried 3,000 feet below the town of Redwater. Where is Redwater, and why does it justify a story? In 1948, it was a mere wide spot in the road - 35 miles northeast of Edmonton and far from the hectic boom of Leduc and the wild Atlantic No.3 well. But, on August 30th of that year, Doug Gamble and crew drilled into the limestone reef and uncovered the biggest Canadian oilfield of them all - 800 million barrels. In a matter of months, this sleepy farming community of 99 souls, mainly Ukrainian, was transformed into a bustling boom town.

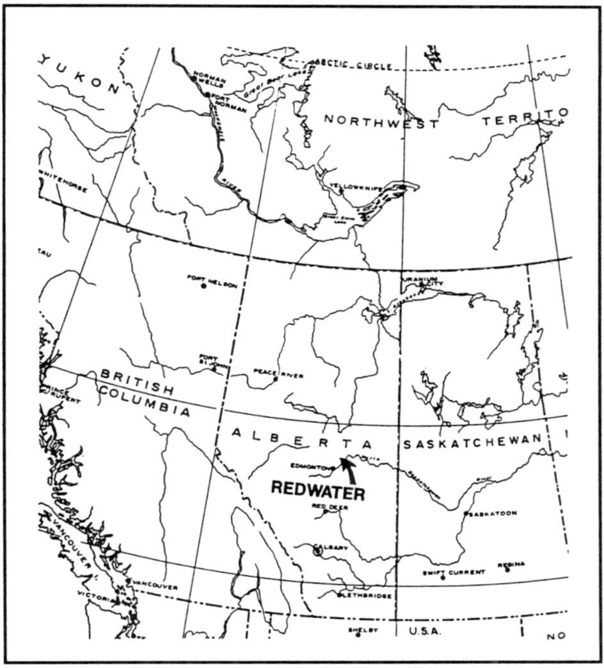

Western Canada Index Map

As with all oil booms, haste was the key word, "Get the oil out of the ground, and send it to market!" The sudden demand for workers was insatiable. Roughnecks, would-be roughnecks, camp followers and promoters descended upon the tiny village, taking over garages and basements for lodging and taxing all facilities far beyond their limits. John Chekerda[1] recalls a bunkhouse that his family had and

1 See Blue Pages for details on the Chekerda family.

the turnover of over 1,000 men. "They came, you know... drifters... only worked for two, three days, and they moved on... We had about 100 beds... double decker army cots... many took away blankets and pillows, and we had to re-stock." The local farmers had reason to be angry too: torn-up fields, spilled drilling fluids, pipelines laid without permission.

An influx of workers from the U.S., completely unfamiliar with Canadian customs, added to the heavy tensions between the oil industry and local townspeople. When the two factions met in the local "watering-hole," the results were predictable, and stories abound of beer parlour skirmishes. Just as predictable was the flight of the drifters once the field was drilled up. Little remains of those frantic boom days: a few pump-jacks nod slowly, 500-barrel tanks now hold crops instead of oil, paved roads have replaced the impassable mud tracks, phones and power instead of party lines and Coleman lamps.

The limestone giant that once held the largest oil reserves in Canada has been drained of all but a few million barrels. It lies, as it were, in its own Intensive Care Unit, with hundreds of miles of "I.V.s" leading out of its body. A few remaining "arteries" pump ever-decreasing amounts of crude from its degenerating heart. Other tubes return unwanted sea water to the reef; the very same water that, millions of years ago, nurtured the living organisms that breathed life into this porous mass.

What of the future? Millions of dollars will have to be spent to abandon and clean up the leases in order to return them, if possible, to their former state. Then, and only then, will the tillers of the soil be able to fully resume their agrarian labours.

Many of those who left the plough to work as oil labourers have returned to farming and can be seen, along with their families, lovingly caring for their land. Their folkloric accounts have not diminished over the years but have grown in dimension and colour to add to the character of Redwater as it is today. Their sentiments are epitomized by Rudy Roloff, an erstwhile roustabout, when he says "I never really left the farm."

Today, Redwater's town park displays the steel derrick which Fin Lineham hauled from Eyot Lake, in the summer of 1948, to drill the discovery well. Larry Chekerda, not born until 1956, realized the historical importance of this derelict mast, rusting away in Imperial Oil's gas plant yard. Single-handed, he persuaded the community and related companies to move and refurbish the derrick. The list of contributors[2] is mute evidence of the successful outcome of Chekerda's persistence and, in every likelihood, Larry will be around to wheel the last valve closed when the field coughs up its last barrel of oil.

Larry Chekerda proudly standing in front of the original discovery derrick relocated in the Town park due mainly to his persistence.

Furnished by Author

2 Appendix I lists contributors.

2 NO OIL IN ALBERTA, EH?

...a structure too big to be believed...

"No oil in Alberta" was the cry, in 1946, from Imperial Oil's headquarters in Toronto. Even its parent's guru, Lewis G. Weeks, referred to western Canada as an over-aged, barren woman with no prospects below the Cretaceous. Who would dare to criticize that erstwhile prestigious firm and its "experts"? After all, Imperial's Saskatchewan campaign had ended the year before in dismal failure - its enormous structures having resulted from differential solution of salt beds that left large mesas as drilling decoys, with nothing in them. Alberta activity was confined to drilling on seismic anomalies only, simply because O. B. Hopkins, Vice President of Exploration, ordered that there be geophysical justification, no matter how vague. Imperial Oil was running out of ideas as to how and where to look for oil!

However, the company's geologic knowledge-workers were not about to give up. Theories ran rampant! "Lewis liked hinge lines. He centered on the Cretaceous as the source of oil," said Wallace Pratt, geologist. Jack Webb and Bill Hancock, also geologists, disputed Lewis Weeks' hinge-line theory that the Western Canada Sedimentary Basin took a sharp dip to form a hinge-like line along which sedimentary sand traps could accumulate and might be oil-bearing. They further proved that this did not, and could not, occur - at least in Canada. However, Weeks redeemed himself as a knowledgeable oil-finder when he discovered the offshore Bass Field in Australia.

Then there was Ted Link, a strong proponent of deep stratigraphic tests, who sent questionnaires in May 1946 to 32 Alberta geological staff to elicit their opinions. Meanwhile, hundreds of miles to the north, at Norman Wells, Oswald Desmond Boggs, friend and colleague of Link, had made a startling discovery while working on the CANOL Project of WWII. Boggs, a Queen's graduate who had studied reefs in South America, had identified the Kee Scarp formation at Norman Wells as a reef, over the doubts of Link and Hopkins. Dr. Charles R. Stelck, University of Alberta professor, in a 1975 lecture on the CANOL Project told of the reaction to Boggs' findings:

> ...I heard rumours, accompanied by loud laughter, that this crazy geologist up there at Norman Wells had gone bushed and thought the reservoir was a coral reef - a coral reef in the Arctic, when everybody knew that coral were confined to the Tropics...

Polar wandering had not been theorized then! Stelck confirmed that neither Link nor Hopkins believed Boggs. They sent Stelck out

---

1 Standard Oil of New Jersey.

with sled dogs to check the Kee Scarp formation outcrop (the Norman Wells producing horizon) for reefal indicators. He returned to state that it was, indeed, coralline, vindicating Boggs.

In the meantime, while this geological bantering was going on, Imperial's "no oil in Alberta" mind-set had spawned plans to construct a Fischer-Tropsch synthetic gasoline plant in Edmonton. The company already had a stranglehold on gasoline retailing and was ready to do anything to retain that position. Plans began under the direction of newly-imported Mike Haider, one of Jersey's few natural gas experts. His mandate: to run Imperial's operations in Western Canada. There seems to be no doubt that Imperial's request for 200,000 acres east, west, and south of Edmonton (Reservation #350), in April 1946, was directly related to finding gas for the proposed plant, as was its geophysical program. Issuance of these concession-type blocks of land required only geological and geophysical commitments.

Ray Walters, born in Gorgona, Canal Zone, Panama, and one of Carter's[2] top geophysicists, was already on the scene, having arrived in 1945 to head up geophysical reconnaissance. (Although it was rumoured in "cowtown"[3] that his real mission was to shut down Imperial's exploration program after three other Carter men had refused this distasteful task.) Peter Bediz, classmate of Ray at the Colorado School of Mines, described him as one who never took anyone's word - he had to satisfy himself by first checking in all other directions. He was certainly a controversial figure, praised by some and vilified by others, but always sure of himself.

Walters first exercised his independent thinking by hiring a Heiland Exploration[4] crew, under party chief Jim Ziegler, to confirm a reconnaissance anomaly in Township 50, Range 26, previously shot by Frank Roberts and crew early in 1946. Confirmation of the anomaly[5] set the wheels in motion to have it tested. The rest is history. Imperial Leduc No.1 was spudded on November 20th, yielded crude on a drillstem test on February 3, and officially went on production February 13, 1947. But it was a modest prelude to the follow-up well,

2 Carter Oil, a research subsidiary of Standard Oil of New Jersey.
3 Calgary, Alberta
4 Jack Armstrong comments: "They had a set of equipment you could pick up the character changes with, and this is why their instrumentation was superior to others."
5 Some thought it was an erosional remnant, others a foothills feature.

"Billion-barrel man." (Ray Walters).
Lantern slide, furnished
by 1946 ASPG Roster

Peter I. Bediz, early 1950s.
Furnished by P. Bediz

Imperial Leduc No.2. Just one and a quarter miles southwest of No.1, this porous Lower Devonian main reef, later to be designated D-3 and still later Leduc, yielded up a black treasure on May 7, 1947, that flowed to the surface in seven minutes - catapulting Alberta into the world of oil!

Little did Imperial realize that yet another chapter in the history of oil in Alberta had already started to unfold. W. N. (Wes) Rabey, Imperial Oil interpreter, who was on the Frank Roberts' reflection seismic crew that initially identified the Leduc anomaly in 1946, recalls the same crew extending their shooting eastward from Leduc and finding yet another anomaly. According to Rabey, it was a very large, prominent and relatively deep structure. However, his interpretation was rejected by the Calgary office, although he presented it several times - it was at odds with their deep structure theories. "How could there be such a thing on the prairies?" The "thing" was, of course, a reef, but they didn't know it then, and neither did anyone else for that matter. Rabey even went so far as to say that there was some attempt to get him to change his maps to correspond to theirs. The story of "Hell no, you can't go" bears repeating when referring to a Carter geologist refusing permission for one of his hands to attend a conference on reefs, stating "Whoever heard of finding oil in a reef?"

Rabey became even more determined to investigate this anomaly further and chose an opportune time, when Frank Roberts was on vacation, to ask the party operator, Bob Grier, how long it would take to re-wire the truck for continuous profiling. "Just a few hours" was Bob's reply. Within a short time, they had shot two miles of continuous profile over what would later be recognized as the east edge of the Redwater reef. The result was astounding - 40 milliseconds of relief! There was, indeed, a large structure - Rabey had proven the Calgary office wrong!

When Roberts returned from his holiday, and Rabey disclosed his "unsanctioned" work, he exploded[6] "You'll get me fired!" The tempest in the teapot soon blew over and, in September 1946, Rabey was transferred to Calgary to do interpretation on several parties' records, in all likelihood because he had "seen" Redwater. He was so enthused about the anomaly that he kept bugging Ray Walters, and others, to file on the land, even threatening to "quit and file on it myself" if they didn't.

"If Imperial doesn't want it, I'll take it." (Wes Rabey)
Furnished by W. Rabey

6 Frank doesn't recall having censured Wes.

Two otherwise unsung groups of doodlebugs played their part in the seismic scramble before Redwater was discovered. Walters, not convinced that it was a deep structure, hired two Heiland crews to do detailed shooting over the feature.

Their task was not an easy one. Road allowances were marked by sharp vee ditches on each side and they were nothing more than trails, many of them not broken out. Swamp areas were covered by 10-ft stringers under which trees had been lain parallel to the road. Trucks had to be winched through a lot of this territory, and the roadways were badly torn up. "It didn't seem muddy enough to stop." This was the cry of the party chiefs when they were being bugged for speeded-up productions - so many days, so many hours. The sand hills were equally formidable. Not only did the crews need to deflate vehicle tires to avoid getting stuck, but records were very poor due to energy absorption. Where it was necessary to enter the farmer's land, off the road allowance, Jim Barron recalls stuccoed mud huts with dirt floors and the occupants being unable to converse in English. They would use their children, who had picked up some English at school, as interpreters. He also remembers some of the shot holes producing artesian water, and yet the municipal authorities didn't bring them to task until the roads were in an advanced state of disrepair.

Imperial decided to drill a test well at Fedorah (known as Redwater "C") location 13-22-57-23 W4, which was on the western edge of the Redwater reef. Wes Rabey had already submitted his recommendation[7] in a memo to Walters, January 1948, which Peter Bediz found "gratifying" in terms of its reference to lithology: "...As early as 1948, Wes Rabey recognized that applied seismology may potentially offer criteria in establishing a relationship between lithology and seismic signatures..."

The well was spudded on March 12, 1948, but was abandoned June 5th after reaching 6500 feet. It had hit the reef 300 ft below what would later be identified as the Redwater water table.

The road from Leduc to Redwater was strewn with dry holes[8] and the oil industry became cautious. Although several companies had encountered the Redwater structure during seismic reconnaissance, they either disbelieved it or were unaware of its meaning. George Grant, recently retired Chevron veteran, recalls his company's crew identifying the structure before the discovery well was drilled.

7 See Appendix II: Redwater "C" Prospect.

8 See Appendix III: Wildcats Spudded Before Redwater Discovery.

Fred Kidd, former Shell geologist, is convinced that the structure showed up when a Heiland crew shot the area for them.[9] Les Clark, also with Shell at that time, confirmed Kidd's statement. However, both Shell and Chevron echoed Imperial's initial reaction that it was just too big to be believed. Despite the Leduc discovery of February 13, 1947, Shell pulled out. One of their senior executives, Fred Davis of Los Angeles, was reported to have said, "Leduc will supply all the oil that Canada needs, so why look for any more large fields?"

As an interesting aside, if air photos and tree types growing along the reef feature had been analyzed, Redwater would have been identified without seismic. Barney Clare, Imperial staff geologist, recalls confirming the outline of the field, visually, during a flight over the area. But who would have selected a wellsite based on creekology? Stories abound in the oil patch of other obvious natural phenomena being overlooked or disregarded, an example being Ida Henkelman's water well in the Leduc area. During threshing time, when Ida's well was taxed to its limits to supply horses and engines, it would start producing gas (which was even lit occasionally).

Shortly after Leduc No.1 went on production, Imperial scrapped the Fischer-Tropsch project and filed on additional open Crown acreage to the east. On April 7, 1947, Imperial applied for another 200,000 acres - Reservation #443, abutting their #350 on the west. Issued in May, #443 would fortuitously encompass the entire Redwater reservoir, but who would know? Edmonton's 1941 regulations, designed to encourage exploration, were still in force allowing acquisition of these concession-type blocks with only geological and geophysical work required. There was no drilling commitment.

Exploration provided a continuous flow of stratigraphic and geophysical information that kept the backroom boys busy. Dr. E. W. Shaw, senior geologist, confirmed in his recommendation[10] to drill Redwater No.1 that the project was shot by geophysical field crews in 1947 and early 1948. Shaw grudgingly refers to seismic data by describing an area of 125 square miles with 150 feet of closure. Had Ray Walters allowed him to see the seismic map by any chance? According to Bud Coote, geophysicist, Shaw and Walters did not get along; both were vying for Jack Webb's job (exploration manager). Walters ultimately won out. Bill Hancock recalls Walters being

9 See Chapter 6 for details re Shell's unbelievable withdrawal.
10 See Appendix IV: Recommendation, Imperial Redwater No.1.

annoyed at Shaw rating a rug in his office and retaliating by decorating his office with an old doghouse carpet.

Although Shaw mentions core-hole drilling in his recommendation, he places little emphasis on it. Gordon Beard's core hole crew covered the Redwater area, a member of which was Roy Murray. Murray, a native of Saskatchewan who had worked on Failing core-hole rigs, considered the Beard crew to be pathfinders in relation to the discovery of Redwater. Shaw also alluded to the large amount of money spent on exploration. In fact, geophysical expenditures for 1946 totalled $620,000. Comparatively speaking, a present-day seismic crew would burn up that whole amount in less than three weeks.

By this time, Ray Walters was pushing to get the Redwater wildcat drilled and, according to Imperial geologist Bill Hancock, Walters took the intiative to work up the final shooting in Jack Webb's absence, stating "If Imperial doesn't want this, I'll take it." (A similar statement had been attributed to Wes Rabey much earlier.) Redwater No.1 was located on Crown Reservation #443 (1-32-57-21 W4M). The drilling recommendation was signed by Webb on May 17, 1948, approved by Mike Haider on June 3rd, and the wellsite surveyed by Jim Carpenter on June 11th with Don Lougheed as rod-man.

One of the first pre-drilling tasks for Bill Smith, Imperial's landman, was to negotiate right of entry, including wellsite, with Hilton Cook, surface owner of SE/32. Kitty (Libbey) Wilkins, a long-time resident of Redwater, referred to Smith as "...a company man. He wasn't out for you, he was out for Imperial." As for Cook, "Gibby" Gibson, District Petroleum Engineer, said he was one of the best leaseholders Imperial had. Gibby found him easy to deal with - Hilton was more concerned with obtaining employment for his three sons, Jud, Jack, and Maurice, than pushing for the highest possible surface rentals. All three were hired! Jud said he was the first to work for Imperial, and that one day he was threshing and pitching bundles when "Fin Lineham came out to the field and took the fork away from me and offered me $1.20 an hour ... I had no idea what the oil business was all about." Jud Cook and his sister, Rita Astle, remember their father visiting the wellsite daily to observe the activity.

Conditions were primitive in those days. Moving the Wilson No.1 rig from Eyot Lake, an abandoned wildcat, had been a horrendous task. Fin Lineham, toolpush, recalls lending a hand along the way: "...I did go out and help them re-timber the bridge where we crossed Redwater Creek ... I brought two or three of the guys ... so the damn drawworks wouldn't fall through..." There were no mud tanks at the

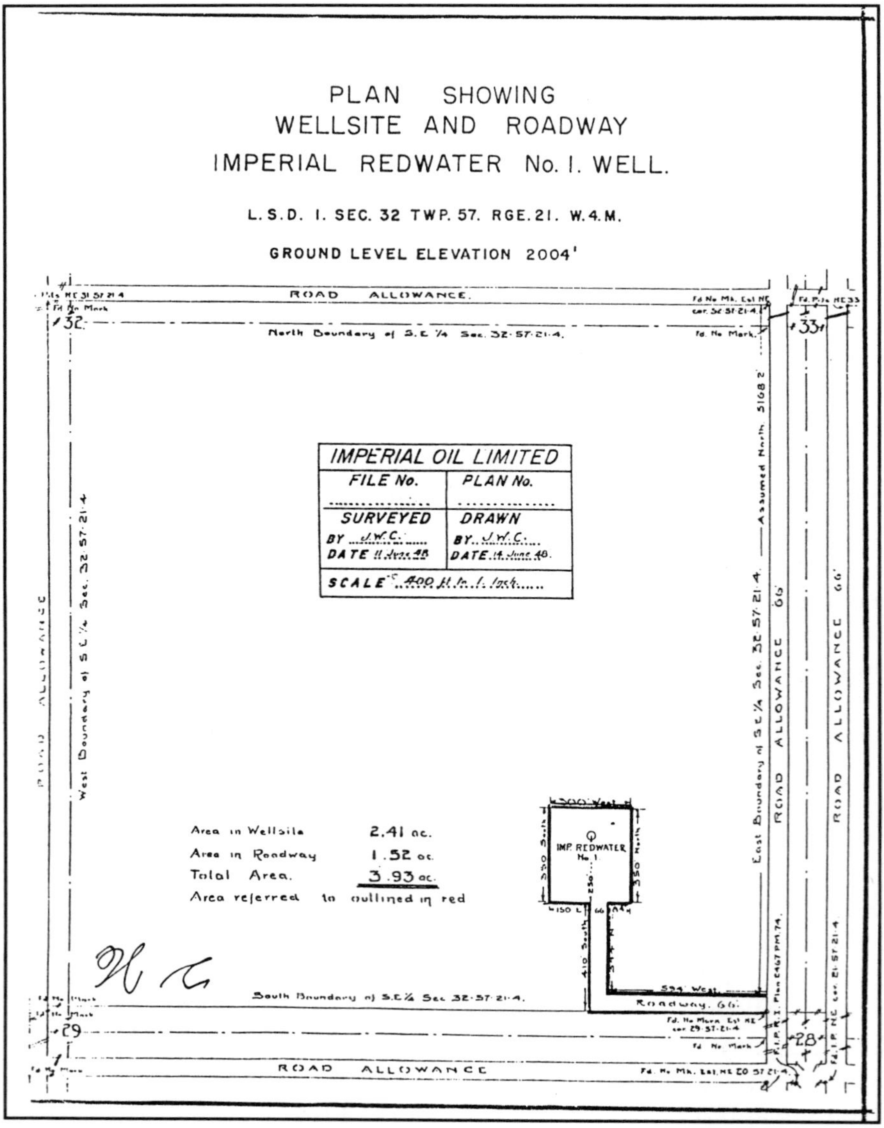

Survey Plan of Wellsite. Note Hilton Cook's initials.

Furnished by R. Astle

wellsite in those days, simply a cellar and suction pits that were filled in. These were considered to be quite solid until one summer's day when Cook was driving a tractor over his land and it sank, leaving only the exhaust pipe showing.

One has to admire the tenacity of the field-workers, many of whom had seen a succession of dry holes and were required to move every two months or so. Rod Morris, the wellsite geologist, had seen his share of abandonments in both Saskatchewan and Alberta and had every reason to believe that Redwater would be just another bust. He hired the Lonsdale brothers and George Kiyooka, among others, as sample catchers. Morris moved into a draughty two-storey house in Opal with his wife, Jean, and worked out of Imperial's chicken coop office in Redwater, formerly occupied by the Royer family's[11] feathered tenants, and rented from Emile Royer, local egg and poultry operator: "...I think I'll rent the chicken house for the Imperial people ... rather than I have chickens in there. They need it bad..."

The Chicken Coop, Imperial Oil's first Redwater exploration office, formerly occupied by Royer's feathered tenants (as it was in 1949).
Furnished by S. Bleakley

Fin Lineham had a portable shack and was accompanied by his wife, Carrie, and their children. One of the drillers, Alex Kenyon, and his family moved in alongside.

Lineham recalls Mike Woodhead, geologist, arriving from the UK and having to go to the train station to pick up Mike's trunk, "It was so heavy, I had to get the railway agent to help me." (Mike collected model trains as a hobby.) Jean Morris remembers Mike's Daimler, a rare vehicle even in the big cities.

Doug Gamble, driller, said Wilbur Anderson was sent ahead to lease enough land to set up camp, including single-men bunkhouses, wash house, tool-push office, and married-men shacks. Anderson was shown acreage at the northeast corner of Redwater, which he could have bought for $50, but laughed at it. "What in the hell would I

11 See Blue Pages.

Author with remnant of Chicken Coop office, (as it was in 1992).
Courtesy of Author

want with this patch of quack grass?" Not long after, the patch was divided into ten lots, at $450 per lot. Again, Wilbur laughed, "Missed the boat."

The well was finally spudded on July 23, 1948 - quite a lag time from when Wes Rabey spotted the anomaly in the spring of 1946. It seems strange that Wes Rabey's drilling recommendation[12] to Ray Walters is dated July 27th, four days after spudding - an afterthought for filing purposes, maybe? (Again of interest in his report are his references to lithology, which further reinforced Peter Bediz's respect for Rabey.) Projected depth was 7000 feet and estimated cost $180,000. Ken Jeffrey, senior accountant for Imperial Oil, solemnly declared on May 19, 1950, that total expenditure was $138,605.24.[13]

As with all wildcat programs, coring and testing of shallower sands was carried out. Both water and gas were found, the latter in sufficient amounts to be used by the community. Everyone was disheartened when salt water was encountered at 2856 feet. The hole inched along, testing and coring revealing "very porous dolomite" at 2929 feet, Green Shale (Ireton) at 2985 feet, and "calcareous green shale" at 3040 feet. Not too promising at this point. Rod Morris's detailed monthly reports[14] reflected this, especially with his most

12 See Appendix V: Redwater "A" Prospect.

13 See Appendix VI for actual breakdown.

14 Doug Lavoie describes Rod Morris, "He knew more about rigs than most toolpushes. Rod's main preoccupation was getting good results from holes."

Helen Kenyon (now Stetsko) preparing breakfast while her daughter, Diane, looks on and husband, Alex (Sandy), laces up his boots in preparation for going on shift at Redwater No.1.

Courtesy of Imperial Oil Ltd.

damning comment: "It would appear as though there is no D-3 here." Ernie Shaw attributed Rod's pessimism to a misinterpretation of the thickness of the shallower beds.

It wasn't until drilling reached 3102 feet that it started to suddenly speed up - encountering porous limestone, oil-stained from 3110-3130 feet. The rig was shut down, waiting for the drillstem test operator. Fin Lineham, in a July 1992 interview with the author and Jean McLaws, said explicit instructions were posted in the doghouse to stop if there was a break in the formation indicated by speeded-up drilling. No one expected this to happen, and drilling wasn't stopped when it should have been. As Lineham noted:

> ...Gamble was 25 feet into it, and I don't know how far he would have gone if I hadn't driven out ... I wasn't gone that long ... he's still drilling away, but he wasn't looking at the samples ... I got the sample catcher

> and I said, "Let's have a look at your samples." ... you could see the oil in it. Why that sample catcher didn't pick it up, I don't know ... So I shut her down, and I wheeled right in, and I got Rod out. Rod was really upset! ... And, of course, I lost my temper. I just told him, "There's a sign up there on the wall that tells what's going on ... Spend more time out here on the rig, then you'll know what's going on..."

Doug Gamble admits that he may have drilled too far and that Fin and Rod did, in fact, shout accusations at each other for not having been around, turning on him afterward. "Slim" Owre, retired senior drilling expert with Imperial, recalls the action taking place on daylight tour and that one of the Lonsdale brothers, both sample-catchers, told Doug Gamble, the driller, that there was something wrong. They shut down and waited for Rod to come out to check the samples.

Recollections of what actually went on at this time are as varied as the individuals involved. With all due respect to them, one has to go to the tour sheets to get the story straight. The tour report shows that drilling stopped on Doug Gamble's graveyard shift on the morning of August 30, 1948, and that most of the day was spent waiting on a drillstem test packer. The Halliburton tester was delayed

Doug Gamble, left, with B.J. Seaman near Weyburn, 1957. "I've just asked B.J. for a raise. He refused. So I'm going to belt him one."

Furnished by D. Gamble

and did not arrive until early evening. Against all standing orders that it should not be done after dark, the tool was run in the hole and opened at 8:01 p.m. A small blow of gas was noted shortly after, dying to zero at 8:20 p.m. The well started flowing at 8:25!

Part of the Drilling Crew[15] at Redwater No.1. L to R, Unknown, Al (Frenchie) Desnoyer, Glenn Gamble, Unknown, Harry (Sledgehammer) Smith, John Taylor (engineer trainee).

Furnished by Mike Woodhead

Doug Gamble claims that a Dowell Inc. tester could not be found, and so Halliburton was called out instead. Doug recalls the Halliburton tester taunting Rod Morris about doing Dowell's work for them, to which Rod retorted, "Go ahead and laugh. This'll be your last goddam test." But Doug's most vivid memory concerns coming on tour at midnight, after they had run the Halliburton test, and finding Don Hunt, short cherubic mud man with an enormous capacity for booze, and Rod Morris "drunker than skunks," Rod having sent Don to town for a case of whiskey. "About 18 stands out (of the hole), she cut loose; however, we had the motors shut down, and she blew down over the pump house roof and filled the mud pit." By daybreak, all the area farmers were there, "lined up like crows on a rail, watching what was going on."

On August 31st, the Toronto office was starting to absorb the momentous news. As Fin Lineham indicated, no one was really prepared for this; there was an unspoken belief that Redwater would be just another dry hole. Even at the wellsite there was a desperate

15 See Appendix VII for list of crew members.

Wilf working derrick at No.1, handling the elevator slips.
Courtesy of Imperial Oil Ltd.

shortage of tubular goods. Seven-inch production casing had to be hauled in from Edmonton, and was immediately run in the hole to total depth and cemented.

The first real look at the producing formation was obtained on September 5th from a five-foot core taken from 3130-3136 feet. Rod described it as "oil-stained porous granular limestone". Not wanting anyone to know what was going on, a prolonged production test using a hook-wall packer was run under cover of darkness. A rate of 19 bbls/hr and a GOR of 149 were recorded. Those present were Maurice Paulson and Vern Hunter, both from Devon, Nate Goodman, Conservation Board engineer, along with Fin Lineham, Fred Killer and Rod Morris. Ed Tovell, Dowell Inc., ran the test. It confirmed good productivity and an under-saturated reservoir.

Shortly after the production test, Jack McCaskill received a phone call from Dave Gustafson, division engineer in Calgary, assigning him as responsible agent in the field[16] pending the arrival of other

16 Confirmed in the Application to Drill form.

Fin with his pick-up truck, 1955.

Furnished by F. Lineham

engineers. McCaskill was stationed at Lloydminster at that time, as petroleum engineer with Imperial. On his way from Lloydminster to Redwater, he remembers seeing a huge pillar of black smoke (Atlantic No.3 wild well burning in the Leduc field).

Following the running of the velocity survey on September 7th, coring was resumed in 5-foot bites, each core being followed by a drillstem test. Rod described the cores as "vuggy." As the hole deepened, recovery improved, getting dense vuggy crystalline limestone. The final core: 3252 to 3254 was noted to be slightly oil-stained, dense to vuggy crystalline, with a washed appearance. The drillstem test interval 3249 to 3254 recovered 2400 feet of fluid, of which 720 was emulsified oil and water and 1680 feet of black sulphur salt water. This test took place on September 24th - nearly three weeks had been spent in evaluating the pay thickness of the section, some 140 feet.

Rod Morris recorded in his diary for September 25th: "Spent most of the day preparing for and meeting Mr. H. H. Hewetson and Imperial Oil directors." This would have been one of the few occasions

Jack McCaskill, first engineer at Redwater, centrifuging a sample taken from one of the drillstem tests to ascertain water cut.

Courtesy of Imperial Oil Ltd.

on which Rod would have to put up with what he referred to as "the Jesus department." A large delegation was to meet with the townspeople and the roughnecks to spread the good word. Walker Taylor, head of the producing department, introduced Henry Hewetson, who said to the gathering: "We came out here to have a look, and we like it. We are happy to be in Redwater." Imperial Oil brought 500 box lunches from Edmonton for the picnic.

On the ensuing days, the crew prepared the well for production. A liner was run and cemented to 3264 feet (FTD) and the well put on production October 7, 1948.

Henry Hewetson, Imperial Oil president, addressing the roughnecks and locals September 25th. Note Lineham's shack in the background.

Furnished by F. Lineham

Walter Hrynchuk and the author beside the discovery marker.

Furnished by Author

Looking back, 1946 was the turning point for Imperial. These two anomalies, Leduc and Redwater, both identified geophysically during 1946 without benefit of geological subsurface trial and error, would ultimately yield over one billion barrels of crude.

NO OIL IN ALBERTA, EH?

3 FOLLOW-UPS

...Now we know we have a reef...
Ray Walters, November 1948

Following the discovery well, interest started to escalate and activity became frantic. Three factors contributed to the spiral of Redwater drilling activity to unprecedented heights:

a) Imperial Oil's need to up production to recover mounting costs;
b) Crown reserve purchasers' desire to recover bonus money as soon as possible;
c) Impetus by both Crown and Freehold lessees to protect themselves from drainage by drilling offsetting crown lands.

Wes Rabey recalls working nights configuring lease blocks that would optimize Imperial's crude reserves. He claims to have tried about 200 different combinations with the help of Fred McKinnon, geologist (who later joined Royalite). Hank Kunst, Ray Walters' assistant, also remembers many alternatives, with final decisions made by Walters. This urgency was generated by regulations, promulgated at the end of March 1948, requiring reservation holders to select leases within 90 days of the deemed completion date of a discovery well, with no wells to be drilled nearer than 4½ miles from the discovery.

By mid 1950, Rabey had decided it was time to leave Imperial. "I knew I was just an employee for a large oil company ... I turned in my resignation to Ray ... He wouldn't accept it ... offered a big raise." But Wes left anyway - he had experienced entrepreneurship with his father, caught the bug, and never looked back. He went into business with Bud Coote, who had left Imperial in January of 1950. Coote's improvements to geophysical instrumentation allowed interpreters to carry the deeper events with more assurance. As for Wes, he can take pride in the fact that he was primarily responsible for identifying Redwater.[1]

Improved reef-finding techniques resulted in two more anomalies being identified east of Redwater in September 1948: Willingdon and Pinedale. It was a bitter disappointment for the geophysicists when these two excellent prospects turned out to be dry holes - erosion of the cap rock had allowed the crude to escape from the 400-foot reef build-ups and be displaced with salt water. However, with the results from contract coverage, the velocity survey at the discovery and its tie-in to the seismic records, Ray was ready to make his next move.

1 Rabey was known as "Mr. Redwater" among his contemporaries.

Imperial Redwater No. 2

Even with hard data, and having a good idea as to where the reef edge was, Ray Walters still needed to confirm its trace. He insisted on deliberately drilling an off-reef dry hole. He selected 3-29-58-21 W4M as the site for Redwater No.2, a little over 4½ miles due north of No.1, to prove his theory. Jack Armstrong, Ray's assistant, confirmed that Ray believed it would be a dry hole and, in order not to be turned down by Toronto office, instructed Armstrong to fabricate a convincing anomaly. This non-existent structure was described in the drilling recommendation as "Seismic anomaly north of the main Redwater trend ... separated from Redwater No.1 by structural low on the seismic picture." The recommendation was signed by J. B. Webb and endorsed by Walker L. Taylor (for J. D. Gustafson) on September 22, 1948. The responsible agent at the well was A. J. (Jack) McCaskill.

Bud Coote described Jack Armstrong, who later rose to the presidency, as a person who would pick out somebody he could trust and then use that individual for information. Supposedly, he was not as good a geophysicist as either Rabey or Coote. An unconfirmed story infers that Armstrong, in the final analysis, was instrumental in having his former boss, Ray Walters, transferred to International Petroleum as general manager stationed in Bogota, Colombia, in 1955. Ray became ill and died that same year - some people said of a broken heart.

Because rigs were scarce, Tip Moroney,[2] production manager, resorted to hiring a crummy Emsco rig from "Dollar for Dick" Harris, head of Can-Tex Drilling Co - the first of many contractors to come to Redwater. The toolpush was S. McDonald, and the three drillers were: Wallace, Forster and Johnson. The well spudded September 28th. The only sign of life was a drillstem test at 4060 ft with a gas blow of 49 mcf/day. A dry hole, as Walters had predicted, confirmed the limits of the huge structure. Jack Harvie recalls Ray Walters' jubilant remark: "Now we know we have a reef." The knowledge gained from this $50,000 dry hole and the indispensable velocity survey would later save millions of dollars by establishing solid seismic subsurface benchmarks and reef tops elsewhere in the field.

---

2 Hero of "Atlantic No.3," he killed this wild well in 1948.

Jack Armstrong in his heyday.
Courtesy of Imperial Oil Ltd.

Imperial Redwater No. 4 (later renamed **Imperial Egremont No. 1)**

Although several wellsites were selected and numbered from 1 to 5, this had no bearing on the priority in which they were drilled. Imperial Redwater No.4, located near the hamlet of Egremont, 3-14-58-22, north of No.1 and west of No.2, was drilled directly after No.2. The drilling recommendation was very straightforward, "Along the same seismic trend as Redwater discovery well." The estimated cost was $65,000.

It was decided to use Fin Lineham's rig again in Egremont. This would be the last of the standard derricks. Jack-knife masts and lighter capacity draw works rendered the big "Wilsons" obsolete (overlong rig moves), especially with the crowded schedules brought on by lease selection requirements.

The well was finally spudded October 25, 1948, hitting the pay slightly deeper than the No.1 well, but indicating a continuous structure (also confirmed by seismic). It was completed December 6th, just in time to assist in lease selections.

Imperial Redwater No. 3 (later renamed **Imperial Amelia[3] No. 1)**

Overlapping No.4 was the drilling of Redwater No.3. It was located 4½ miles southeast of the discovery well in 9-11-57-21 W4. The shortage of company rigs resulted in Imperial going with a newly formed outfit called Canada West Drilling, with Charlie Hawke as

3 Named after Hilton Cook's mother who, with her husband, Samuel, formed the first settlement: Cookville in the 1890s, south and east of Redwater.

the toolpush. The well seems to have been more in the hands of the engineers; the drilling recommendation stated "test as required by engineers". The licence to drill was signed by J. D. Gustafson, okayed by D. P. Goodall, and ratified by N. E. Tanner October 4, 1948. Jack McCaskill was once again the responsible agent in the field.

At about this time, G. T. MacCallum joined Imperial's office in Redwater, providing comprehensive weekly geological reports. He reported Redwater No.3 as "Completed on January 3, 1949 at 3303 with an IP of 1935 bpd after 16,500 gallons of acid had been run."

Imperial Redwater No. 5

Imperial Oil discovered a loophole in the regulations which allowed step-outs closer than 4½ miles if they were located on freehold property. Such a location was available in the NE/25-57-22 of Imperial's Reservation #128, over two miles to the west of Redwater No.1. Mineral rights were held by Canadian Pacific Railway, and surface rights by Sophia Hrynchuk.

As with No.2, No.5 was drilled with the major intent of gleaning information. The recommendation for drilling was vague: "...flank location on Redwater structure ... unknown subsurface structure ... detailed seismic coverage." In a letter to Mike Haider, Jack Webb indicated that the hole should be drilled as soon as possible, and so another relatively new contractor was chosen: Northern Development. (Walter Hrynchuk, one of Sophie's sons, remembers Paul Guthrie as the toolpush.) Imperial's letter to Northern dated November 20, 1948, specified that the well be drilled with "due diligence and perform such drilling at the request and in accordance with instructions of our representatives." It would give them $600 per day for daywork. Interestingly, the letter does not quote the footage rate. It was authored by D. E. (Ed) Lewis for Walker Taylor's signature. As with Redwater 2 and 3, Jack McCaskill was the responsible agent in the field.

No tour sheets are available. The well was abandoned January 11, 1949, after about six weeks of drilling and testing. This west step-out helped delineate the productive limits of the reef by disclosing the fact that the limestone was back-reef and therefore virtually non-porous. Fin Lineham remembers Fred Killer telling him why No.5 was no good, "Because, right at that particular point - like a straight line ... It's got curves in it, like this - that's the way these formations that we're drilling in are. And they were on a portion of where there was no water laying when the ocean dried up in the first place." Figure that one out!

Seismic activity was unprecedented, attracting some innovative geophysicists with unusual ideas, for example Wentworth Tellington.

Wentworth Tellington sent the author a brochure (English on one side and Spanish on the other) outlining his diverse career. Born in 1916 in Gorham, New Hampshire, he traces his mother's forebears back to the founders of Dartmouth College, sea captains, and even secret service to kings. Academically, he has covered four institutions of higher learning and even more subjects, including geology and geophysics, and is reputed to have been a West Point graduate - having served there as an instructor. His publications are equally diverse and voluminous.

Went, reminiscing over his unbelievable career.
Furnished by W. Tellington

Tellington's story reads like a Horatio Alger dime novel. He started in Oklahoma's oil patch. He found his way up to Calgary via Century Geophysical shooting Amerada's Olds acreage. According to his write-up he became chief geophysicist for Pacific Petroleums Ltd. by 1950, having been hired by Lorne Falconer and Art Nauss. "Lorne and Art thought they wanted an innovative geophysicist, so they hired me." They later regretted their decision re Wentworth because, as Frank Ronaghan recalls, "He cost us a lot of money."

Jim Ziegler,[4] then with Century Geophysical, recalls him in most unflattering terms, "You couldn't supervise the guy. He always knew better" The following are excerpts from Went's correspondence with the author indicative of his unorthodox approach to life.

> ...We had two seismographs[5] competing over exactly the same ground using six 60-foot spreads. I had two eager computer guys. (Could one of those have been Louis Castelli?) I told them to look at the 24-trace seismic records and make believe they were detectives looking for anything unusual ... Within three weeks we had turned up positive but vague irregularities in some of the east-west lines ... It seemed at first like some kind of background noise ... Then this phenomenon traced on our contour maps began to curve, following a more or less northwest to southeast trend which matched nothing we could find on the surface ... I told everyone to forget it because it was just some dumb problem we would have to live with, so the whole staff went back to plotting the usual three or four horizons. After discussing it with Art and Lorne in great secrecy, we decided that the reef formations might cause phase changes in the regular waves bouncing off that Devonian formation.
>
> We did some mathematic doodling with the data from the geophones. We compared this to the raw stuff, did some isopach mapping to check for compaction ... We thought we could plot the west edge of the Redwater field and relate that to the salt water line. So we went for broke and, in the next several months with Scotty Tosh calling the shots with the drillers, we stuck our necks out and picked 52 drill sites just on the edge of the fantastic, high bidding areas where we could buy a bargain; 51 of those wells came in and made Frank rich...

Went remembers a speech he made to the SEG in Dallas. Wes Rabey recalls sitting there like the cat that had swallowed the canary when Tellington revealed what he thought was a great secret, but Rabey and his cohorts already knew all about the outline of the reef.

Ziegler recalls that "When Tellington was with Bear Oil, he invented a seismic train that was supposed to be transportable over the muskeg ... couldn't move it ten feet. It was made so light that it all pulled apart." In his March 22, 1993 letter, Tellington refutes the claim that his inventions were failures.

---

4 Jim Ziegler said he had been asked to replace Wentworth, but he turned it down.

5 Gordon Hess states that he and his C.E.C. party did all the shooting for Bear and Pacific. Tellington merely looked at their records.

> ...The super-light vehicles and aluminum buildings were used again and again by us at Pacific Pete ... We had bunkhouses, offices, drill and dynamite vehicles, all basically on special-built bombardier tracks ... They were free rolling, about 24 inches wide, mounted on two-axle assemblies ... The buildings were designed by me and an engineer out at the old (Edmonton municipal) airport ... We had a problem with the sides buckling, because we made them too rigid ... I chose very light and flexible, and it worked - a light aluminum sheet with a panel of flexible insulation glued to it ... We learned to make them not more than 24 feet long, 8 feet high, and 8 feet wide. And we could pull them like a train, with two or three hitched together with ordinary light trailer hitches. The floors were ½-inch plywood...

He is now the proprietor of Deep Moon Gold Mine in Downieville, California. His brochure mentions "42 years of married life with four different and exciting women ... two children and five grandchildren." He refers to himself as "not a wealthy man in the ordinary sense of the word." He has given away more than most people earn in a lifetime, but he is "wealthy beyond imagining in the long and solidly made number of fellow seekers."

During the author's research among the Calgary doodlebugs George Hobson emerged as a geophysicist of some significance; with his Master's from University of Toronto, he was the first Canadian party chief for Heiland. He arrived too late to shoot Redwater, but he worked Waskatenau. This is where he met Ed Lambert, who had signed up with Heiland as a jug hustler.

Drilling continued at a frantic pace. Somehow, Bud Kelly, Imperial's drilling superintendent, managed to keep some semblance of order as he oversaw and examined the growing number of daily drilling reports. Shell had missed the boat by pulling out and allowing its Redwater option to lapse, but other companies were more astute.

Anglo-Canadian and Home joined forces to purchase the E/2 36-57-22 on November 10, 1948 in the first crown reserve sale[6] after the discovery well. Redwater No. 16-36 was spudded just a few weeks later on December 3rd. All told Anglo-Home drilled eight holes on their half section - 26 trucks hauled crude 24 hours a day. Gord Connell, Royalite engineer, believes they were producing 2000 bpd per well. (No question about it: Eric Harvie's Section 31, immediately offset to the east, was being drained!) Royalite, newly liberated from

6 See Appendix XVIII listing all sales by legal description, successful bidder, and bonus amount in dollars as compiled by H.H. Somerville, also Chapter 7 for more details.

Imperial and with over $6 million burning a hole in their pocket, purchased Sec 15-57-21 on February 15, 1949. Fred McKinnon recalls Gordon Connell taking ballots to come up with the price and getting an agreement on the average: $3,154,000.[7] In a hurry to recoup their money, they hired not one but four rigs, each on its own quarter.

The mad search for drilling contractors was continuous, perpetrating a phenomenal influx of outfits from the U.S. and the creation of new Canadian firms. Labour was difficult to come by, many newly-arrived immigrants were hired from the Commonwealth and Europe. Some started out in menial jobs but were promoted rapidly into positions of higher responsibility, sometimes even to their own surprise.

Work conditions were often nightmarish: unskilled workers, language and cultural barriers, frigid temperatures, lack of equipment and supplies, and inadequate housing. Turnover was at a high level, especially among workers from the U.S. who were unfamiliar with Canadian drilling practices and climatic conditions (there were many who had never seen snow).

R.H.J. (Bob) Elliott, an Imperial geologist at Redwater from fall of 1949 to spring of 1950, had a few tales to tell about frigid conditions.

R. H. J. (Bob) Elliott and his trusty panel truck.
Furnished by S. Bleakley

7 See Chapter 7 for more details on Royalite's purchase.

Elliott routinely froze up his car but, more often than not, managed to get antifreeze into the carburetor before it quit. With the approach of winter, he recalls all the fellows wanting garages. However, by the end of the season, cars were left outside to prevent condensation in the gas tanks.

Elliott also recalls a Can-Tex crew working their day shift right through to midnight because the afternoon shift never showed up. When the graveyard crew didn't make it out to the lease, they concluded that enough was enough and abandoned the rig. It froze solid! How long it took old "Dollar for Dick" to thaw it out is anybody's guess.

Another incident involved a derrickman, obviously green, who foolishly wore rubber boots in -40° weather. When he came down from the mast, the driller observed the boots with alarm and said, "How are your feet?" The fellow said, "Pretty cold for a while, but they feel fine now." The driller ordered him to remove them immediately. Elliott was there when they uncovered his feet - they were white. Poor chap lost some toes - a lesson to those who wore rubber boots.

However, many stayed and many more came - the oil patch hummed with activity. It was an era of risk-taking, industriousness and entrepreneurship, abounding with tales of hardship, hair-raising incidents and colourful characters. The following illustrate a few, but represent many:

"Kayo" Blackwell was rumoured to be a full-blooded Cherokee. Hugh MacKenzie, retired service rig operator, has vivid recollections of this character who conducted outrageous manoeuvres and, in so doing, obtained much notoriety.

> ...He was pushing for Parker Drilling ... and he moved up to Redwater.. I was out at the lease when they were rigging out.. after the first well. And he always had a cigar in his mouth - always unlit - he chewed them. I was talking to Kayo - I had just released his rig from the completion. And he had an American driller - one on each shift. And they never changed shifts. If you were the graveyard driller, you were the lowest one on the totem pole...
>
> Kayo got a bonus for every day he knocked off what they figured for on that well. And he was poking 'em down as fast as anybody. If you wanted to haul water for Kayo, you'd give back 10% of your take ... He'd buy timbers or planks, and he wanted so much back from the lumber yard on the price - but always in cash...

On this particular occasion, the driller was lowering the derrick, and the saddle broke on the hook, and down came that derrick! Just in a complete mess! And I happened to be looking over at Kayo ... and I seen the cigar just disappearing down his throat. He ran over there ... I was still in my truck doing some paper work ... He came over, and he said, "Say, did you see my cigar?" I said, "Yeah, I seen it all right. You swallowed it!" But, that was the first well in Redwater that he drilled on...

Anytime you were working on Kayo's rig - I would be overseeing the swabbing on it sometimes - you had to watch that he didn't have all the catwalks already tore out behind you...things like that. Because, the minute he was released, he wanted to have the trucks within 10 minutes to start moving things... I can see him yet, hollering at them roughnecks. They'd be carrying pipe across the yard... a joint on each hand with the two guys... "I don't want two joints to two men, I want two joints per man going across there." Oh, he was a ... slave driver! And he was just a little, short guy, too... about 5 foot 4, and tougher than whalebone, that little devil...

MacKenzie describes rig moves.

...Athey Wagons or huge toboggans made out of steel plates. And they'd just leave 'em underneath and hook a cat on each toboggan, and away they'd go ... Some of the wetter areas ... they'd just skid it on those old toboggans...

I can remember, one night, we were completing a well. And we'd had a chicksan break up in the derrick while we were acidizing. Well, there was no acid ... we were just using the flush oil to clean up the lines, and this chicksan broke pretty near the crown of that derrick, and it just covered that whole rig in oil. So I ... went over there and told Kayo (to) douse all the lights on the rig. And I was back ... checking one of my other crews, and I looked back over ... there's the lights back on that rig. I went back over and I finally caught up to Kayo and I said, "Hey them lights are on again." So Kayo went back over ... and out went the lights. About midnight, I looked over there, and the lights were on. So, we went over, and Kayo walked up to the driller and asked, "What's the lights on for?" He said "Well, we have to have lights to work." Kayo said, "No. You have orders ... Don't have any lights on." And he grabbed the derrick hatchet off the tool board in the doghouse and went half-way down between the light plant and the rig and just put that derrick hatchet right through the main power cable and said to me, "Now them lights will stay out tonight for sure." He was just one of those proverbial characters...

> At first, Kayo had all-American crews, but they couldn't cut the weather up here. They wouldn't show up on shift. They'd be gone ... catch a bus back to Oklahoma.. Oh, we had an awful time that first winter - winter of '49-50. Oh, that was cold - 62 below one morning in Redwater. I can remember Kayo came into the office that morning, and Abzinger was taking the reports from him - Julius never did say too much. One of the other clerks was sitting there and Kayo said, "Boy, it's cold out there." The clerk said, "Well, it don't matter after it gets 40 below." Kayo said, "Well, if it don't matter, what did they put all them numbers on that temperature gauge for?"

Alfred Edwin (Ed) Barroll has come a long way from the many miles of roughnecking to senior manager with Mobil and now a hard-nosed, uncompromising leader as Chairman of Petro-Can. Stories of early days illustrate how he succeeded.

Born in Calgary, the son of a butcher, he served with the RCAF in WWII, took one year at Mount Royal College and then obtained his degree at the University of Oklahoma. Upon graduating, he received two job offers from Gulf and a third from a drilling contractor, T.T. Eason, who had just received a contract with Imperial Oil in Canada. He signed on with the latter as a trainee-driller and recalls roughnecking on Eason's Emsco and Brewster N4 rigs at Redwater. The two toolpushers and the drillers were all U.S. citizens and, as far as Barroll was concerned, incompetent.

T.T. himself was a highly respected hardware merchant who got into iron and then decided to venture into the drilling business. His son, T.W., was a drunk - at least that's how he appeared every time he came to Edmonton. The only person in the Edmonton office (10237 - 98 Street) with whom Barroll had any contact was Cloyce W. Ryland, nicknamed "Goat". He was supposed to be the head honcho, but Barroll termed him an idiot. Louise, wife of Ray Tull, thinks nepotism played a part in his appointment. Goat had previously supervised rigs in the Illinois area. Eason's Redwater maintenance was virtually non-existent with equipment constantly breaking down.

Suddenly in December, heading into the worst winter the area would ever experience (1949-50), the two toolpushers took off - having forgotten their coats in Texas. Several of the drillers headed back and, finally, they all left. This set the stage for Barroll to be set up as tool-push on one of the rigs and Reg Nelson, formerly of Can-Tex, to push on the other. This in the face of -50°F weather. And the commuting of the crews out of Edmonton meant virtually 12-hour shifts, because it was two hours to drive out and two hours to drive back in. Many mornings the roughnecks, locally hired, would not be at their appointed corners in Edmonton for pick-up. The other roughnecks - some of them having served time - while driving past Fort Saskatchewan wished they were back behind those bars. At least it would be warm. Ed remembers the frozen sandwiches, the impossibility of keeping warm, the burning up of body energy with no chance of any fat accumulation, "A helluva way to get fit." But there was no question that Ed not only hardened, but also learned how to manage men - a true leader with a stern resolve to get the job done.

Ed's only contact with Imperial Oil was when he took in his time sheets, and Bud Kelly would have his red pencil ready, "We're not gonna pay you for that. That's rig repair." This is probably where Ed honed his bargaining skills. He had no idea of the contract or the influence of R. B. Curran, newly appointed Imperial Oil manager, who no doubt had brought his buddy's (T. T. Eason's) rigs up and put them to work in spite of Tip Moroney's protestations. Brutal, fierce working conditions, and yet there were a few faithful workers. Ed describes his condition, "as hard as a rail, a real test of character." He roomed in a large boarding house in Edmonton.

Goat's hiring of Metro Moholitany to be a general gopher-nipplechaser- chauffeur backfired on him and turned out to be one method by which the Ukrainians got back at the 'you-alls'. Metro made a career out of screwing Eason by stealing gasoline and shortchanging them on many occasions, but Goat was unable to trip him up. They appeared to hate each other's guts, but there was, in Ed's words, a residual affection of sorts. Another story re Goat: One summer day the crew were dying of thirst - they had no water -and who should fly over the rig but a bit salesman and Goat as a passenger. As they swooped down over the rig, Goat opened the door and yelled, "Get with it, you bastards." If nothing else, it taught Ed the stern story of how things ought not to be done and how lack of management created problems. There is the story of how they were going to set surface pipe and somehow the table bearing fell down the hole. After a great deal of excavating and cribbing, one of the crew was lowered on a rope in his derrickman's harness - finally hooking on and retrieving it. By that time, Ed had had about 48 hours of non-stop work and took off for a well-earned rest.

"Some time in March or April, a delegation of Socony-Vacuum (later Mobil) people came out to interview me at the rig with a job offer." He'd had enough of drilling, so he accepted. From there on he progressed steadily through the Mobil organization to one of the top jobs in Canada.

Ray Tull was born an Oklahoma farm boy in 1924. After attending country school as far as Grade VIII, he made his first career decision at age 13.

> ...We were raising horses and mules at that time, and Dad said, "We haven't any money to send you to high school.. If you want to go, we'll make a way to go. Or I'll give you a team of mules. I said, "I'll take the team of mules and start farming...

He later considered the oil patch when he was told, "On a drilling rig, you can get 99¢ an hour, time and a half, and double time."

> ...So I went down to a rig ... about five or six miles from the home place ... I was out there at the end of the catwalk, looking around.. The pusher drove up and says, "A big, husky guy like you should be working on this rig." I said, "That's what I'm looking for." Three days later, the phone rang to go to work. That was for Falcon Seaboard Drilling Company from Tulsa, Oklahoma. And I roughnecked for them at 99¢ an hour ... We worked straight seven days a week ... and Falcon Seaboard was having trouble getting men. I doubled (shift) anywhere from three - sometimes five - days a week...
>
> One day the pusher said, "The office tells me we're going to send two rigs to Canada. They're thinking about.. going up there to look around and do some bidding.. Would you consider going?" I said, "Well, whatever you think." Jerry Stauss was vice-president, and he's the one I talked to. He said, "Ted Law and Chet Nestor ... say it's an older man's place - with more experience. I think it's a young man's, and I want you to go." That was probably August or early September. That was the last I heard of it until about the middle of April '49. I called a report in, and Jerry said, "Have you got your red flannel drawers? You're going to need 'em. You're going to Canada." I said, "I got to go home first ... and talk about it."
>
> Louise and I were married in November of '44. We also had two daughters by that time -one was born in January of '46, the other in December of '48...

He recalls the move very well.

> ...I let the trucks get a one-day start on me ... I knew I'd catch them at the border. I crossed it on the 6th of May, '49 ... And I'd get up to Edmonton and find a place to unload the rigs. They promised that after I got things unloaded, I could come and pick up the wife ... drive up my own personal car with her and the kids. But things developed

quicker ... I got up there ... got everything organized and found a place to live in Edmonton. But by the time I got my doghouses and water tanks built ... ready to go, I didn't have time to get back for the family because they started the rigs up. So my father drove them up. I sent him back on the train...

Then they told me I had to move into Redwater for Canadian Gulf Oil Corporation. We rigged up just on the edge of town, about a quarter of a mile away ... I was toolpush on both rigs. They worked me hard ... I spent more time in the car than I did at home. I probably averaged two nights a week, in Edmonton, in bed. We had a shack for one of the rigs ... but I spent most of the time between the rigs ... back and forth. If I wasn't running casing on one, I'd be running tubing on the other ... on the go all the time...

Ray recalls the two young geologists sent to sit on the wells, Rob Laurence and Neil Taylor:

...Then they took a twelve-hole contract with Gulf for the two rigs to drill - straight day work.. $660 a day. I told them it was too low. But they said they could make pretty good money in Oklahoma at that rate. I told them, this is not Oklahoma...

He confirms the difficulty in getting and keeping workers.

...They let me bring six drillers and six derrick men ... three for each rig. And we hired the helpers up here. We all came in on a six-months work permit. I believe we had one driller and one derrick man left before we even finished the first hole ... We wound up getting a few farm boys, and that's where we got our best men. You could depend on them. I was lucky. I went to the University at Edmonton to see if anybody was interested in summer work, and I got four or five boys there, which really worked out well for us. They were eager, and they wanted to work...

Lanky Desjardins came to our door in Edmonton - Louise couldn't understand him ... I took him out to the rig ... And he had four or five more French buddies. They all came from north and west of Edmonton. Lanky worked for Northern Development when they drilled Imperial Redwater No.5...

When winter really hit, a few more moved away. At the very end, after the first winter was over, we only had five guys left of the original group...

When asked about his experiences with the locals, Ray fondly remembers Mr. Bigoray, one of the surface owners on whose land he drilled. Bigoray would invite him into the house every so often to have a glass of home-made wine.

"The Troc" night club, 103 Street, Edmonton, 1951. Only pop bottles to be seen; booze was hidden in handbags and under the table On the left, Ray Tull and his wife, Louise and A. L. (Lanky) Desjardins with Madeleine, his bride.

Furnished by R. Tull

After a successful business career checkerboarded by ups and downs, Tull and his wife, Louise, are now happily retired. An all-season Canadian sportsman, Ray is into championship curling and low handicap golfing. His den provides ample evidence of his prowess in these two sports: badges, cups, medals and plaques.

It was Tull's drilling skills which readied Gulf's freehold reserves for production, but Paul Evans' engineering know-how would send Gulf's oil to market.

Paul C. Evans was born in Illinois, but his family moved to Texas when he was an infant. He went to high school in Fort Worth, then enrolled in Texas A&M University in 1933. Being an All-State basketball player, he received a scholarship which gave him enough funds to pay for his room and board - he reckoned the total cost of his four years to be $1600.

Evans describes the primitive stage of petroleum engineering in the early thirties with courses at different universities, trying to come to grips with new scientific principles. When asked, for instance, whether the concept of super-compressibility was understood in those days, Paul facetiously answered, "We didn't even know the word."

> ...My teacher at A&M, in the graduating class in Petroleum Engineering, selected Gulf Oil Corporation to be my employer. It was the Gypsy Division out of Tulsa, Oklahoma." Evans signed on in 1937, along with 34 other green engineers, for the enormous salary of $125 a month and was trained in every division of the company. He recalls, "There was a strong feeling among the field hands, and much of the management, that we engineers were going to take everybody's job...

His only experience in oil and gas, up to that time, was in his senior year - a two-week stint in east Texas to work for an employee of the Texas Railroad Commission (our equivalent of the Alberta Conservation Board). Paul was obliged to see that the pressure was holding and the cement was setting up properly on the surface pipe jobs. "So, I reported to the commissioner every morning at the pool hall. For 14 days I watched them play pool, and the people bringing in the charts, and he would sign his name to them."

On December 26, 1937, Paul married Sarah-Beth. "I had just started to work, and they didn't look very fondly on my getting married. They asked, 'You won't get married while you're on the training program?' I said, 'No, no way'." Paul remembers meeting Sarah-Beth in October, having one date with her, and shortly after buying the rings.

Apart from his Army service (Dec 7, 1941-1945), Paul was with the Gypsy Division of Gulf Oil until the early part of 1949, when Eddie Gallagher walked into Paul's office and said, "Somebody's going to have to go to Canada." Paul replied, "Well, it's not going to be me. It's too far away. My wife is expecting a baby." Later on, Gallagher said to Paul, "I know you don't like this, but I've got to do it. The boss wants to see you in Tulsa tomorrow morning ... I've already got your ticket" and handed it to Paul. Evans said, "I've been demoted into going up to Canada. There's nothing I know about it but igloos." He discussed it with Sarah-Beth over the weekend. Sarah-Beth thought that was a great idea. "We could go to Canada for a short time, and when we had a long weekend, living in Edmonton, we could drive up to Alaska."

Paul recalls arriving in Redwater and Gulf Oil wheels coming up to see the operations. His first exposure to Canadian football made him think that the Canadians were very conservative, "As they went along, every time the third down came up, they punted. That's real conservatism, to punt on the third down" (fourth down in American rules).

When asked if he made many trips to Redwater, he replied:

...Practically daily, especially in the winter of '49-'50 ...We used to lay flow lines, used to need four crews out there - one working and the other three warming up. And they'd work 15 minutes at a time ... The steel pipe, if you hit it with a hammer, would just come apart ... There was nothing unusual in seeing a man working as a toolpush for a drilling contractor within months in Canada; whereas, down in the States, a toolpusher would usually have 15 to 20 years' experience before he was set up...

Paul Evans, vice president , Oilfield Technical Society, 1951.
Courtesy of Canadian Oil & Gas Industries, Nov 1950.

Paul was involved in organizing the first Oil Show - at Devon in 1950. "We had a one-day meeting. Jack Godwin had a tent and some refreshments in the back room ... During the day, I travelled heavily in through Jack Godwin's place." He also recalls the excellent write-up given them by the old Edmonton Bulletin, "What a wonderful show it was, and it was going to beat the Tulsa Oil Show ... Speaking of Godwin, Jack was the living end when it came to barbecues, he was known as 'Mr. Barbecue'." Paul was one of his assistants on these occasions.

Evans moved on in 1951 to a company called West Canadian Petroleums owned and operated by French coal interests. It was during this time that he came into contact with Harry Webster, a person for whom he has high praise, and eventually bought out.

Paul remembers Canadian Export Gas being organized at Oyster Night at the Petroleum Club with Wilbur Griffith and two others, the idea being to export gas to the east. One of the first people Wilbur hired to head up the company and to run the exploration arm was Gus Beck. "Then, in July, he hired me to head up Production. Several years later, Newmont came in ... Gus had died in 1974. I was made president. In 1976, Placer Development of Vancouver moved in and purchased all the shares, the company becoming Placer CEGO ... They let us run the company, kept all of our employees. We expanded in money and numbers ... That $55 million investment got to be about three quarters (of a billion). A pretty good rate of return."

Paul and Sarah-Beth became citizens in 1965. "I think I was very fortunate to come to Canada, to stay here. The country was real good to me, and I think we gave a little bit back ... We enjoyed it here ... saw the best of the oil business."

McIntyre & Webster, drilling contractors. Harry Webster started his career, as had many others, in Turner Valley in the '30s, working on one of the last derricks to be hacked out of wood. He worked his way up through Royalite Oil, with Charlie Visser as friend, boss and mentor. With the slow-down of drilling in the Valley, Webster went wildcatting on the plains with the Imperial Oil parent. It was he who rigged up the Cinderella rig (Franks No. 3) at Grassy Lake No. 3 in the fall of 1945 which subsequently drilled the Imperial Leduc No.2 discovery well in May of 1947. He is one of the original few who never worked in Leduc.

Webster was pushing tools up at Morinville when he heard from Boyd McIntyre, a long-time friend, that Royalite was looking for a drilling contractor. They decided to become partners and pursue this opportunity. They learned that Cardwell had an outfit, not exactly what they wanted, but their sales representative, Len Walker,[8] persuaded them to take the TR-57 Trailerig, 5,000-ft capacity (pulled doubles powered by two D-13,000 Caterpillar diesels) - the first Trailerig to operate in Canada. Webster received assurance from Charlie Visser, "If it doesn't work out, you have a job waiting for you." With no down payment required, Webster and McIntyre had a nice kick-start into the business.

Webster said he was unable to recruit many hands from his former employer, "Anyone working for Imperial thought it would be too risky, working for us." He managed to hire some from Commonwealth

---

8 Started as a roughneck in Turner Valley in 1936. After a three-year stint in Trinidad with an engineering firm, he returned to Canada to work for supply companies joining Cardwell in January 1949 as Canadian representative.

Drilling and several from Turner Valley. One of his key men was Lynn Donaldson, experienced and steady.

Donaldson developed a high regard for Webster and feels his motto could very well have been, "It's not what the oil industry can do for you, it's what you can do for the oil industry" (to paraphrase John F. Kennedy). He describes Harry "as a builder, not a character, straightforward, and not the type of guy to attract stories." Harry was a humble man with whom many people enjoyed working and doing business.

McIntyre & Webster were assigned Royalite's SE quarter. The other three quarters each had their own contractor: Commonwealth Drilling, Canuck, and Archie Miller. The first hole to be drilled was located on 8-15-57-21 W4. It spudded March 22, 1949, and was completed April 8, 17 days later, at 3168 ft. No. 1-15 started April 15 and completed April 29. No. 2-15 spudded May 5 and was completed May 17; and No. 8-15, May 20-June 1, each hole requiring fewer days as they got the hang of it.

Road ban was on when they were drilling their holes, but they were fortunate in being alongside Redwater Creek, making it unnecessary to haul water. A steam pump brought water to the rig. The crew trucks managed to get in and out of the lease most of the time. Service rigs were an unknown quantity at first, so the drilling crew ran tubing, swabbed the wells and put them on production.

Other jobs followed in rapid succession. Webster said that both Imperial and Royalite kept them busy for three years in and out of Redwater. In the spring of 1952 they purchased a second Cardwell rig, this time a Model "O" with a capacity of 8000 feet.

By the end of 1952 their reputation for reliability was established, so it was no surprise when Paul Evans and two other would-be purchasers approached them. One of the latter suitors pulled out his cheque book, but a tentative letter agreement had already been signed with Paul Evans, representing a three-company partnership, Canadian Fina Oil Ltd., West Canadian Petroleums and British Empire Oil. Paul Evans, then-manager of West Canadian Petroleums, describes Webster, "...A wonderful individual ... an honest man. His word was his honour, and no one could do anything but admire him. He'd never do anything to hurt anybody ... only helped everybody. I had a wonderful relationship with him ... an unbelievable man." Webster was retained to manage his old firm and Boyd retired.

The Seaman brothers subsequently bought out Webster in September 1958, who left in April 1959 after making sure of a smooth

L. to R: Harry Webster, Len Walker (note Hughes bit watch fob), Cardwell rep., and Boyd McIntyre standing before their new Model TR-57 "Trailerig" on the Royalite lease (SE/15-57-21 W4M), Redwater, spring 1949.

Furnished by H. Webster

transition to the Saskatchewan brothers. Later in 1959, he became Drilling Superintendent for Canadian Fina and was given many important assignments in Canada and abroad. He retired from their service December 31, 1969.

"Hughie" Leiper was a toolpush before he had reached 21! Not surprising really, when one thinks of the sudden, insatiable demand for roughnecks (most of them green as grass) in the Redwater of 1949. If you were any good at all, you were drilling within three months. Besides, he had some guidance from Sandy Addison, who was a push with General Petroleums.

Leiper had started off in Turner Valley, roughnecking weekends during WWII. He first hit the oil boom at Leduc in July of 1947 on General Petroleum's steam rig. It later collapsed and met a fiery end, falling into the Atlantic No. 3 crater on September 6, 1948. His first encounter with Redwater began on Texaco's Crown Reserve parcel purchased at the July 15, 1949 Crown Reserve sale. Gordon Savage

Group of some McIntyre & Webster roughnecks: L to R, Bill Rhining, Bob Brundage, Heinz Zwingel, Merle Batdorf, Joe Machovec, Ralph Archibald, Bob Melny, Rudy Zwingel.
Names verified by Harold McIntyre, Red Deer (no relation to Boyd or Gordon).

was the company representative and Frank Flewelling was his push. He and his bride, Irene, bought a 7' x 12' trailer with no insulation - 1949-50 was the coldest winter they ever experienced with frost on the inside walls.

Hugh and Irene Leiper standing in front of their General Petroleums pick-up at a Redwater wellsite.
Furnished by H. Leiper

By March 1950 Hughie was set up as push, and he now had Frank Flewelling as his competitor. Both were using National T-25s, which pulled doubles. Cody Spencer, General Pete's boss - always looking for ways to speed up operations, offered the pushes a brand new Stetson hat if they could move their rig within four hours from the time the operator released the rig until they spudded on the next location. This took a lot of preparation at both ends. The Company Rep. would help by deducting two hours on the sly for the time of the well completion while either Athey Wagons or aeroplane tires were brought in under the rig ready to be jacked up with the derrick still erect. The new location, of course, had been completed with mud pits, matting and timbers, ready to take on the rig. Needless to say, Hugh had several hats in his wardrobe.

One incident which nearly cost him his job happened in the dead of winter; they thought they could skate the rig down one slope and up the other over a small coulee. It got away from them as they were easing the outfit down the slope and ended up rocking back and forth at the bottom; fortunately, there was no damage. The moving trucks were then brought around to the other side of the coulee and the rig was winched up. The drillstem tests, in Hugh's view, were exceptionally hazardous. He wonders why there were no fires because, when the drill pipe unloaded oil, it would spray up through the mast over the hot diesel engines.

Another gimmick to 'hurry' (Cody's favourite word) operations was covering up the "Waiting On Cement" time, which was supposed to be 24 hours: cement samples would be caught and put in cans over the exhaust to dry them out, as if they were observing the 24-hour "WOC". The Conservation Board never checked, but Imperial did come around in an attempt to keep them honest. Three fellows that Leiper remembers fondly are Lornie Leeson, Lou Glendinning, and Bill McGreevy - all under the jurisdiction of Bud Kelly, drilling superintendent.

Hugh reckons that his average time from spud to release was 5-6 days, and he recalls drilling about 50 holes in the northwest part of the field. Art Branscombe, Charlie Visser's nephew, reported in an OTS Drilling Symposium "...that the wells are drilled to about 3300 ft with about seven bits in 10-11 days and a few record breaking 7-day efforts with little or no lost circulation problems." Hugh isn't sure whether it was he or Frank, but one of them had the record for the fastest 8-hour tour: 1300 feet in what Leiper called a snow bank.

Quoting from the Edmonton Journal of Wednesday, June 21, 1950, Flewelling's crew "...are setting a blistering pace, latest record being 44¾ hours to drill out from under the surface casing to the top of the D-3." Flewelling was born in Lacombe, got his start roughnecking in Turner Valley, and marked the Joseph Lake Viking oil discovery as the high point of his career. Part of the crew consisted of Clifford

Reid, Dudley Moore, Sheron Baker, Ernie Chapot, W. Shepard, Jack Rosichuk, Emil Proust, Donald Moore (brother of Dudley), Roy Emslie, L. (Jim) Mutch, Floyd Snortland, Jim McMaster, Allan Henderson, A. Makowsky and Hank Newfeld. Jim Mutch remembers his summer work as a roughneck during his undergraduate days. He obtained his BSc in 1953 and joined Mobil. After many years with that firm, he now has a managerial position with PetroCan.

Hugh and Irene were promoted to a 12' x 18' shack - Ralph Binning, one of the senior executives, making sure there was an addition to make it more like a home. Hugh recalls driving to Imperial Oil's office every morning to hand his reports in to Bud Kelly. One morning he had picked up a hitchhiker who was working as a summer student for Imperial and was very pleased with the money he was making. Hughie lashed out at him - told him to get back to school. The chap heeded his warning and became Number Two at Bill McGregor's company, Nu-Mac. His name was Don Baker. Great Plains later beckoned Hughie to the Drumheller area. His experiences with Glen Phillips, witcher, south of Munson and the Drumheller Syndicate generate a set of stories of their own, to be told another time.

In 1957 Pacific Petroleums hired him as superintendent in the Fort. St. John area to replace Scotty Tosh, who had wanted to retire. By March 1961 he was production manager of the Alberta area stationed in Edmonton. It was during this time that he revisited the famous Rebus Quarter, where he had roughnecked on Atlantic No.3, but with a different mission: Persuade farmer John Rebus to sign his royalty interest unitization document. In a similar fashion Hugh would now visit Redwater wearing an entirely different hat. He remembers entering the shut-down district office with Kelly Gibson and pitching out hundreds of production records. The din from those many rigs was silenced by the time Leiper returned to Redwater to look after Pacific's production. That, too, diminished with proportionately higher water output. In 1978, PetroCan took over Pacific's and other companies' wells.

Hugh is presently Executive Vice President of Guard Resources but still takes time out to reminisce about the old days, nearly 50 years of "turning to the right" behind him.

Bob Dixon. Most recently (early 1993) a rank outsider, Penn West, threw its hat in the ring when it acquired 167 wells from Petrocan for an undisclosed sum. The ram-rod behind this transaction is R. K. (Bob) Dixon, a 1950 graduate of Oklahoma and C.E.O. of Penn West. Dixon also started off in Redwater right after leaving school, being hired by Bert Corey of New Superior. A veteran of several companies (Fina, Merland, and Opinac), Dixon is extremely optimistic: "There is still a lot of oil out there." This is supported by the fact that 65 of the wells he bought are producing about 900 b/d net oil with 95% water.

Penn West, with ambitious work-over plans, may even increase oil production. Who knows? Redwater, already on its last legs, may still have another ten years of life.

Bob Dixon in his office.
Furnished by Author

Garnet Edwards started in Turner Valley and is a veteran of the Atlantic No.3 blowout. He drilled two dry holes for adjacent Leduc Consolidated. He was then loaned to Commonwealth Drilling in Redwater and, for a while, was "swing push" under Al Howes. He recalls a Cardwell RL rig, all mechanical, all dog clutches - no hydraulic, being moved up to the Opal area in -40°F weather at about 5:00 p.m. The mast was on the ground, but Al wanted to get the derrick up and ready to spud-in first thing in the morning. He told Garnet to start pulling on it, using the A-frame, but Garnet wouldn't touch it - saying there was going to be trouble. He then turned to the other push, Herschel Keyser, to start reefing on it. As soon as a strain was put on the A-frame, the cold caused the brittle steel to shatter into a thousand pieces. The welder needed several days to repair it.

Garnet also recalls one of the Parker toolpushes, known as "Kayo" Blackwell, who was going to show the Canadians how to make hole. He recalls Kayo just barely drilling out from under his surface casing when Edwards was at TD (total depth) ready to run long string.

Tom Wark recognized that drilling was one thing but moving rigs was quite another; he could see an opportunity here and partnered up with Pat Patterson to form a moving contracting firm. Time was of the essence in those days, and Wark's outfit had this innovative procedure down to a fine art. He claims to have moved the McIntyre & Webster rig in 55 minutes and could take on jobs that no one else would touch. They were getting $500 per move in Redwater.

Wark describes the procedure.

> ...The minute the rig was released, there was a knock-out arrangement around the wellhead so that the jacking-up of the rig onto the tires coincided with the hook-up at the front of the tractor. When the rig, with everything in the derrick - including the travelling blocks and the kelly - stood over the new wellsite, they hooked up the shock hose from the pumps to the standpipe. And they had actually started drilling dead center before the jacks were slacked off and the wheels had been taken out from underneath. That was how I did it in 55 minutes...

Part of Tom Wark's career was spent with General Petroleums, Drilling Contractors. During this period, he experienced a bizarre situation. One of the many hazards to which cattle were exposed was "pipe dope". This oily red-lead compound was slathered by the roughnecks on the "boxes" of the drill pipe to ensure efficient spinning in (making up) of the threaded tapered joints when running in the hole. For some reason, this compound had an irresistible, fatal attraction for cattle. A few licks of this red poison and they were dead.

Tom Wark recalls a lease which had not been properly fenced before the rig moved in. The crews had already trucked in the drill pipe, which had plenty of red lead on the exposed threads. The farmer was soon confronted with approximately ten dead animals, the cause of which was first evidenced by their bleeding tongues, which had been cut on the sharp threads of the drill pipe. He took the stomachs to the veterinary in Edmonton, who confirmed that the cows had, indeed, been poisoned. One cow was very highly prized by the farmer's wife. Asked what he wanted in settlement for that particular animal, the farmer didn't know. Because his wife had arthritic hands, his main concern was that the replacement have soft enough teats so

Tom Wark in right foreground. Note the 16 aeroplane tires (from bombers), eight on each side, each with hydraulic lifts, and the shackles that hook up the two sets of tires to a common yoke hauled by a single tractor (five tractors needed under wet conditions).

Furnished by H. Webster

that his wife could milk it with minimum difficulty. General Petroleums, drilling contractor, had to settle the claim, which included two unborn calves.

Elmer Gavel, a native of Olds currently weakened by heart problems, still retains much of his old vigour when he recounts his bit-tipping career - an expert in the field. Elmer apprenticed in the Valley as a welder under Lou Archibald and obtained his Journeyman ticket.

Operating conditions were hazardous: The "borium"[9] tipping and hardening substance used to reinforce drilling bits emitted poisonous gases, and fumes from the acetylene torches were inhaled as there were neither ventilators nor gas mask. Lou's son, Ralph, who later became a safety expert, is still bitter over the treatment his father received. "All he got was a $100 cheque and a certificate on the wall." His pension was only $75 a month. Ralph believes that his dad's ingestion of borium fumes hastened his death.

Elmer Gavel was introduced to the world of wildcats when he was sent up to be mechanic on the Imperial Coalspur No.1 deep foothills wildcat drilled in 1944. The location had been selected from Hank Kunst's surface geology in the spring of 1943. (The author was rod-man.) After reaching the Mississippian, truckloads of mud loaded with lead were brought in to build up the hydrostatic pressure so as to prevent blow-outs. "The drill cuttings (samples) boiled in the water pail." (entrained gas). Elmer recalls them bailing out the hole and lighting a flare about 3-4 feet high. Tight carbonate - "lots of pressure, but no capacity". It was a steam rig with six boilers, and W. J. "Gibby" Gibson was the wellsite geologist. He shared his shack with Elmer, and they became very good friends.

Elmer had married Evelyn McIntyre in 1940. Evelyn comes from a Turner Valley oil family, her brothers working for different companies, including Home Oil. When Fin Lineham's Wilson rig was drilling Eyot Lake deep test (south of Leduc) the family rented part of a garage at Millet for $10 a month. Evelyn recalls getting the word to move the rig to Redwater. "We were on the move to Redwater ... We tied two chairs together, collapsed the table, and were on our way."

Upon arrival at Redwater from Eyot Lake, Elmer and Evelyn rented the front of a house - the back half to Desrosier (truck driver) and family - at $8 a month from a Mr. Horbal, who ran a store next door.

When drilling started, Evelyn recalls asking Elmer when he came in from day shift, "Have you got a dry hole yet?" Elmer's reply: "No,

---

9 Elmer describes borium as crushed tungsten and carbide in a hollow steel tube.

L to R, Gavel and Gibby in front of their bunkhouse at Imperial Coalspur No.1, 1944.
Furnished by E. Gavel

but Rod is sure worked up ... (laughter)." Elmer continues, "Rod had great hopes for it. Then it looked like it was a fizzle." He claims that Rod told the driller to put on a fresh bit and drill to total depth, run a cement plug, and abandon it. "That very night, they lost circulation ... couldn't keep up with the mud ... put everything we had into the mud, couldn't control it ... gassed up the hole so bad ... mud was so light, it blew. And them spies were all over the place ... later on, grabbing leases on the basis of the information." Elmer corroborates the drilling into the D-3, "We didn't core into it. We just walked into it, drilled with a bit ... The next day, it blew mud all over the rig."

Evelyn remembers the excitement when they hit oil on the drillstem test, "Rod got so drunk ... They had an old car ... the four doors were all open, the rad was leaking, but.. went to town." Elmer remembers Hilton Cook buying a case of whiskey, and everyone was drunk before they got to the beer parlour, whooping and hollering. Elmer recalls, "I flaked out ... out cold. But there was a party, with the wives, that night." When asked about the reaction of the townspeople, Evelyn said, "They couldn't comprehend ... It was a shock to them."

Elmer now could see a way of capitalizing on his welding expertise, and asked Fin Lineham's advice about quitting and setting up shop, tipping bits. With Fin's okay, he bought a truck with an electric arc welder on the back of it, on credit, and paid it out in a matter of months. His expertise expanded, "We were the only ones that could make them stick using bare rods," referring to the protective sleeves on the 'subs' used with drillstems. Elmer's expert touch in remodelling the Hughes OSC3 bits was legendary. Even the Hughes salesmen brought bits to his shop to be cut down and retipped. Contractors were now pounding on his door, because they were getting 600 more feet of hole out of Elmer's modifications. The regular item usually made only 1600 feet and took two hours longer.

As is sometimes the case, there was a snake in the grass. A person who shall remain nameless snitched on Elmer. That individual went to the Hughes office in Houston and told them, "There's a farmer-welder up there remodelling your bits." Hughes lost no time in modifying their OSC3 to an OSC3A. On top of that, Elmer's truck driver went to work for Hughes. But all was not lost - the big 13-inch and 15-inch surface bits still needed constant retipping. He also got into farm implements. He later sold out to his brother-in-law and son.

Summarizing their lives, Elmer says "Live and let live." Evelyn: "Life's been good ... We've had our ups and downs. We've been blessed with seven grandchildren." Thus ended one of the best interviews that I have had.

PLAT dated 1951

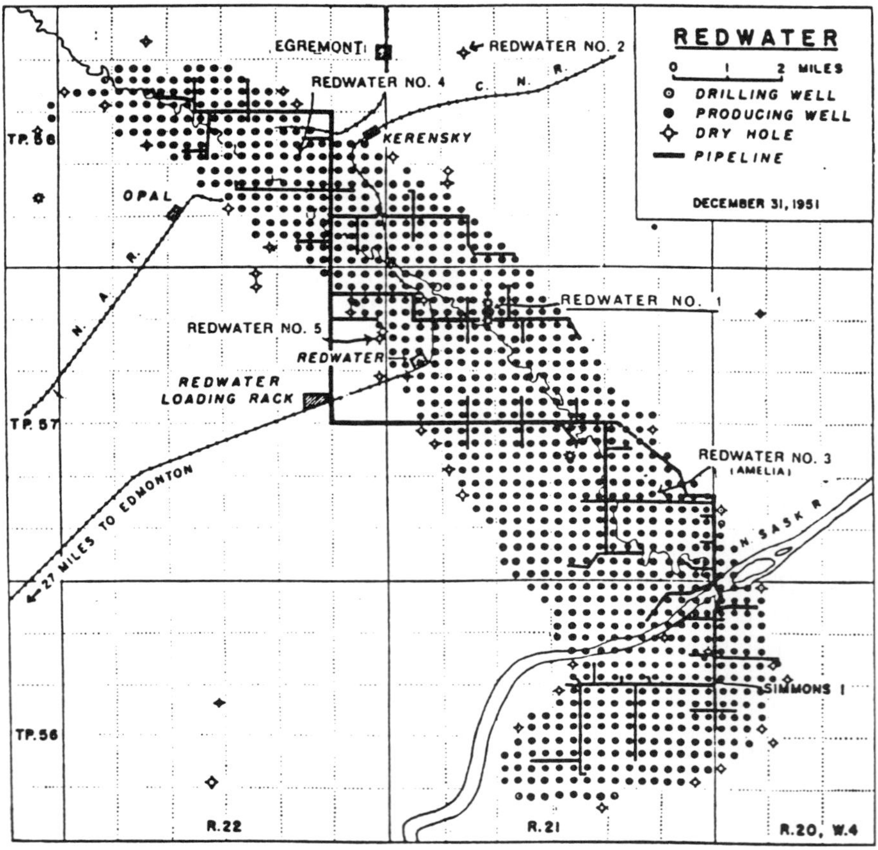

from "Dynamic Decade"

Looking back on it all, over 1000 holes had been drilled by countless hundreds of workers in a little more than three and a half years, yielding not only a wealth of oil but a wealth of information for the geologist and the engineer. . .

4 FOR THE GEOLOGIST

(and those intrepid souls who are willing to venture into the arcane world of Geology)

The Mysteries Unravelled

The discovery of Devonian "coral" reef oil at Leduc, in 1947, ushered in a brand new era of thinking for the petroleum geologist. Up to this time, these earth scientists had shown no interest in coral reef as a possible target for oil exploration. The exception was Des Boggs, the "crazy geologist", who had identified Devonian reef at Norman Wells while working on the WWII CANOL project. Gradually, after many analytical studies, the significance of these structures became known, throwing off the shackles of conventional wisdom epitomized by: "no oil below the Mississippian".

Their findings not only took them forward into new theories, such as "polar wandering", but also into the past theories of renowned scientists such as Charles Darwin. His landmark works[1] on the evolution of coral reefs provided a wealth of information, the earliest and most prominent being "The Structure and Distribution of Coral Reefs" published in 1842. This invaluable guide soon became a "best-seller." Of interest, too, were W. C. Krumbein's WWII studies of coral reefs in the South Pacific, which assessed military beach landing capabilities. Although the majority of these studies were conducted in the Pacific, it is important to realize that this was once the type of environment that existed in Western Canada within which the Redwater reef had its beginnings in warm, sub-tropical seas.

Canada, Calgary in particular, became a centre of excellence for earth sciences, the need to know more about reefs being driven by an industry desperately searching for Devonian reservoirs. In the first four years, over 20% of the known proven reserves of Alberta were discovered in six giants in the Edmonton area: Leduc in 1947, Redwater in 1948, Golden Spike in 1949, Acheson in 1950, Wizard Lake and Bonnie Glen in 1951. While looking for new ones, geologists were also studying these monsters in detail in order to obtain answers as to their producibility and suitability to waterflooding or other forms of pressure maintenance.

Although many companies jealously guarded their geological secrets and allowed no one to see their geophysical records (except on a trade basis), a few did permit staff to publish their findings, for example, Jim Law's landmark paper on the Devonian of northwest Alberta written in the early 50s (which Chevron later regretted okaying). The Geological Survey of Canada had no such restraints - their watchword being "publish or perish." Unfortunately, the time lapse between writing and appearing in print was measured in years.

---

1 See Appendix VIII (Darwin bibliography).

Even so, Dr. (Ralph the Rover) Rutherford[2] castigated himself and the Geological Survey of Canada for not having had the wit to extrapolate reef outcrops in the mountains east into the subsurface, where their counterparts were waiting to be discovered.

Why was so much more geological effort applied to Redwater than to Leduc? Both are bioherms, reef build-ups on the Cooking Lake platform. Leduc had been subjected to alteration of rock type, being dolomitized,[3] but had relatively uniform lithology throughout. The fact that the D-3 oil pay was originally 38 feet with a gas cap on top and a water table below, one moving down and the other upward with production, ultimately resulted in a five-foot gas-oil-water sandwich when the pool was nearing exhaustion. This was entirely an engineering matter, and none of the companies really "explored" Leduc except for edge plays.

By contrast, Redwater is a build-up of pure limestone secreted by marine organisms which extracted calcium carbonate dissolved in the sea water. These communities, or colonies, developed skeletal frameworks which acted as baffles and binders. They have an algal content plus cements and internal sediments - one of the chief reef builders was the stromatoporoid, believed to be somewhere between animal and vegetable. A. J. (Jack) Goodman, of Socony-Vacuum (later Mobil), favoured lime-secreting algae as the main reef builders in a rather vague report dated November 5, 1948. Although given wide and immediate distribution among Imperial's Calgary staff, in retrospect it did not really make much of a contribution.

The Redwater reef, in particular, is an atoll type "bioconstructed" and not aligned on any obvious trend, but with a rim beyond the productive limit to the southwest. It has fore-reef (seaward east side), main reef, and back-reef (west side) types of limestone and these comprise zones representing different growing conditions in the life of the reef. The reasons for examining these in detail, and understanding the environment under which they grew, were economic. Each of these zones has its own unique rock types and was studied by oil company geologists with a view to estimating the oil reservoir capacity of each. These earth scientists also relied on micropaleontologists, such as Imperial's Dr. Diane Loranger. As a

2 Then University of Alberta professor of Geology who, prior to Leduc, had vowed to drink all the oil "West of the Fifth".

3 Dolomite is the double salt of calcium and magnesium carbonate.

microfossil expert, she deduced the environment in which certain forms thrived and, from that the rock type and position in the colony.

Several knowledge-workers stand out for their academic contributions to a better understanding of the geology of Redwater:

John Andrichuk was born 50 miles to the east of the town of Redwater to Ukrainian parents. He was awarded a PhD from Northwestern University, Chicago in 1951, one of his professors being the widely respected Krumbein. His early years were spent with Gulf under the tutelage of Max Littlefield, expert in carbonate geology. While working for Gulf, he was given the opportunity to analyze the "Green Hornet Report"[4] secretly passed to him by John Downing. Appendix IX comprises Andrichuk's original longhand notes on this study, dated April 1953, indicating his excellent grasp of the Redwater reef.

He envisioned a rim around the edge of the reef similar to the atolls of the South Pacific - an inverted "saucer" with more porosity from the rim outward than underneath the central depression. Subsequent tilting of this saucer upward toward the northeast would allow oil to migrate towards its lower rim, where the water drive would also be more effective. He also knew that the northeast edge had been subjected to storms and wave action.

Andrichuk published a monumental treatise in the Bulletin of the American Association of Petroleum Geologists, January 1958. In it, he described the prolonged growth of reefs in the Redwater area:

> ...Vertical carbonate development is about 1100 feet. Over this mass lies the shale known as the Ireton, or Green Shale, which is up to 125 feet in thickness. The types of limestone and their reservoir capacity vary laterally and vertically to such an extent that only general zonation is observable. After the deposition of the massive reef, crustal movements shifted the reef vertically, creating fractures. In addition, the settling of different parts of the reef, with the increased sedimentary load of some 2500 feet of shales and sands, resulted in further alterations...

Even today, he feels that there are patch reefs south and west of the former productive limits of Redwater similar to those that grow in other back-reef backwater atolls that could yield hydrocarbons. This is a challenge for the budding geologist.

By 1966, Andrichuk had moved to Alex McCoy, a U.S.-based consulting firm with offices in Calgary. Another associate was Ralph

---

4 See Chapter 5 for details of this report on Redwater's reservoir parameters by the Redwater Working Committee.

Edie, former Shell geologist. They soon decided, "If McCoy can do it, so can we," so they formed a partnership which, unlike that of so many other geologists and geophysicists, has stood the test of time. They still operate out of one room, no secretary, one phone, no answering service, no Fax, no word processor. If the phone doesn't answer, they aren't in - it's that simple!

Andrichuk & Edie, a durable partnership of two top knowledge workers
Photos by Bryan Rutley.

J. E. (Ed) Klovan was born in Rich Valley, Alberta, north of Gunn and Onoway and west of Edmonton. He grew up in Fallis, attending an eight-grade, one-room school, the highlight of his educational career. He wouldn't have missed it for the world. It was during his time at Fallis that he was influenced by a geophysical crew to pursue earth sciences as a profession. The crew would ride around in his dog sled and he, in turn, was allowed into the recording trucks to learn how geophysics worked. Some of the crew even helped him with his homework.

His university life started at Edmonton, Alberta, under a solid, well-backed faculty: Folinsbee, Stelck, Warren and others. His summer employment was with Cal Standard, working in Jasper on a surface party, then switching to helicopters and reconnoitering the Rockies all the way up to the Arctic Ocean. He graduated in 1957.

Ed Klovan holding his Brunton to obtain the dip of an outcrop, circa 1963.
Furnished by Ed Klovan

Ed Klovan was greatly influenced by Andrichuk's studies. John, in turn, encouraged Ed, who was proceeding toward his PhD at Columbia U (NYC). He credits Andrichuk & Edie with influencing him to proceed with his thesis on the Redwater reef. Some detractors said there was no point in studying it because there were few fossils, but he proved them wrong. Embry and Newell, senior geologists at the G.S.C., thought he should find a reef close to home, and that's why he chose Redwater. At that time, Columbia University was in its heyday with such greats as Marshall Kay and Maurice Ewing.

Klovan completed his thesis in 1963. His dissertation for his PhD appeared in the Canadian Society of Petroleum Geologists (CSPG) Reprint Series No.1 as "Facies Analysis of the Redwater Reef Complex."

Klovan approached the subject of the Redwater reef with several questions designed to invigorate his fellow earth scientists:

1. Is the carbonate structure at Redwater a reef in the sense of modern geological thinking?
2. What role did organisms play in the growth and development of the reef?
3. How did the structure develop through time?

Klovan describes a reef as:

> ...A rocky elevation or knoll on which there is depth of six fathoms or less at low water ... Structures must show evidence that they were erected and strengthened by organisms capable of building and raising a skeletal framework above the sea bottom and potentially capable of

growth into turbulent surface waters to which they would be wave resistant...

This leads on to his technical definitions of various parts of the reef, such as "inclined fore-reef deposits". He also points out the presence of limestone mounds in the North Sea currently at depths of 1000 feet and in temperatures of around 5°C. These masses attain heights of 200 feet and cover areas up to 16 square miles, so warm water may not have been all that essential.

Grant Mossop, now Director of the Institute of Sedimentary and Petroleum Geology, was Ed Klovan's pupil. Mossop agreed that Redwater was an atoll mechanism with the interspersing of shale and dense lime. His contribution is equally important.

Reginald Alexander Shatford could be said to be the Christopher Columbus of Redwater. Born in Vernon, B.C., in 1901, he received his secondary education at King Edward High in Vancouver and his BSc in mining engineering from McGill in 1925. We have no record of his employment until Red Lake in northern Ontario.

Shatford's graduation photo from McGill 1925 yearbook.
Furnished by McGill Archives

One story has him visiting Calgary, from Vancouver, and being interviewed by Les Clark of Seaboard. He was turned down and went over to Imperial. Ernie Shaw, Imperial's senior geologist, recognized

Reg as a mature administrator and hired him as Rod Morris's replacement, as District Geologist. He would oversee the unruly young geologists at Redwater despite his lack of knowledge of soft-rock geology. When Redwater was drilled up, Shatford moved into the Edmonton office. Rumours state that he had ruffled feathers by pushing too hard for his (Precambrian?) theories, resulting in his being blackballed. This could have made it extremely difficult to get another job in Alberta in his chosen profession, but it apparently did not.

From 1954-59 we find Reg in Calgary, employed by Compagnie Francaise des Petroles as a researcher. He had been hospitalized and was recovering when he received his pay cheque plus a pink slip. He was successful in fighting this and obtaining a settlement. He returned to Winnipeg, and his drinking increased. Not long after, he was killed in an automobile accident.

Reg Shatford and George MacCallum taking it easy in the chicken coop.
Furnished by S. Bleakley

Shatford applied himself to studying and writing up the geology of the reef.[5] His reports remain as a memorial to his research efforts, done in the field and typed by Lorraine Fox. A.D.M. (Tony) Mason describes his style as turgid. The author thinks his writings leaned toward the theoretical. Jean McLaws, researcher, says his scholarly manner is hard to follow. This may be because he and his staff were still groping for answers. Reg assigned projects to his geologists: Hec MacRae to a cross section, Doug Lavoie to the suitability of rock facies to accommodate waterflood, and he himself studied the implications of the Precambrian surface as a control for the positioning of Redwater and other reefs. Being a hard-rock man, he went to great lengths to relate fracture patterns to existing drainage and fields.

5 Shatford and Andrichuk liaised with informal discussions.. When Reg presented a paper (at an ASPG luncheon?), he unwittingly failed to acknowledge John's contributions. Reg rectified this by writing Andrichuk a three-page letter of apology.

His Progress Report #1, written in February 1951, was the outcome of a January 15th organizational meeting. Judging from the remarks of those present[6], it was obvious that his main job would be the selection of water disposal wellsites. He had also been instructed to study "...pore-size distribution as a first step to understand the manner in which reefs were formed."

Shatford restored the reef to its original position by tilting the reef back, 40 feet to the mile. This tilting, in Reg's view, "...merely compensated for regional tilt, but failed to fit many of the local facts." He created a cross-section:

> ...Restoring the reef, on the assumption that at the end of Ireton (Green Shale) time, the top of the Green Shale would be essentially flat and that subsequent compaction of these shales might be approximately one-third of their present thicknesses ... The very major portion of the distortion took place between the end of Devonian and the beginning of Cretaceous times or that time represented by the great unconformity...

Shortly after, he asked Hec MacRae to do a cross-section to see if there was any possibility of internal correlation. The micrologs gave evidence of porous sections while the radioactive logs, in MacRae's opinion, "...provide a quantitative measure of porosity, but their value in lateral facies correlation is not there."

In a later report, dated August 25, 1951, Shatford deals with excess salt water production, noted almost from the beginning and now exacerbating. Because the porosity and permeability decreases from the east side toward the west, the force of the water drive might be expected to be most effective around the reef rim rather than in a northeasterly direction. The Redwater reef runs many miles west and south towards the Morinville reef (see plat). There are billions of barrels of water downdip which provide an aquifer. The northeast edge of the reef is marked by a talus slope consisting of broken fossil fragments. This northeast edge was exposed to wind and wave action from the Devonian seas. Storms, from time to time, would very likely break up the reef front and create this talus slope which at times may have even been exposed to the air.

He attributes the "making of water on the west side of the field" to a partial local raising of the general oil/water interface due to permeability barriers. This is true because the porosity is not as good.

---

6 Vernon Taylor, M. P. Paulson, D. B. Layer, J. W. Young, R. D. Sluzar, A. S. McLatchie, and R. A. Shatford.

On the other hand, the sector with the good porosity actually started to water out sooner and allowed the poorer porosity on the west to continue to produce.

Turning to photogeology[7] Shatford's study states: "It is not too farfetched to expect the present surface to reflect the Precambrian surface ... Lines of structural weakness form a pattern, usually northeast and southwest ... Tectonic effects might tend to be a controlling factor as to where the reefs might grow." He then goes into some detail about surge channels and channelways, "The explanation for these phenomena cannot be confined to sedimentation and reef growth but must involve tectonics."

Reg Shatford, with hard rock always in his subconscious, was quick to read the signals at Redwater, but he stretched his ideas beyond the realms of credulity to further his belief that Precambrian basement shifting controlled the accumulation of all the oil in the region. He, therefore, delivered himself of "A Preliminary Treatise with Respect to Tectonic Influences on the Redwater Reservoir." It opens with a swipe at sedimentary geologists: "It wasn't until about 1949 that progressive and competent sedimentary geologists began to see it (creekology?) as an important factor ... It does not receive the consideration it will demand in the solution of our exploration and production problems." This semi-technical, but nevertheless important, aspect of oil finding could have resulted in the discovery of both Leduc and Redwater using the surface phenomena alone.

In his summary, "...Full cognizance be taken of the fact that the Precambrian basement is not static, has not been during the period of geosynclinal development and probably never will be ... The Precambrian pattern ... might be a consideration in locating undiscovered fields." Finally, "An understanding of tectonic influences on the Redwater reservoir may become necessary in solving our production problems" (a rather cryptic note).

A. D. (Andy) Baillie is a world renowned earth scientist capable of communicating effectively with any group, regardless of language or race, at the same time not "talking down to them." This dedicated 81-year old mentor lives his profession, teaching being his avocation. He cautioned the author that there can be confusion between reefs built by stromatoporoids and those built by algae. He suggests the

7 The author's examination of the mosaic and discussions with his colleagues convince him that even the reefal zones have surface expressions.

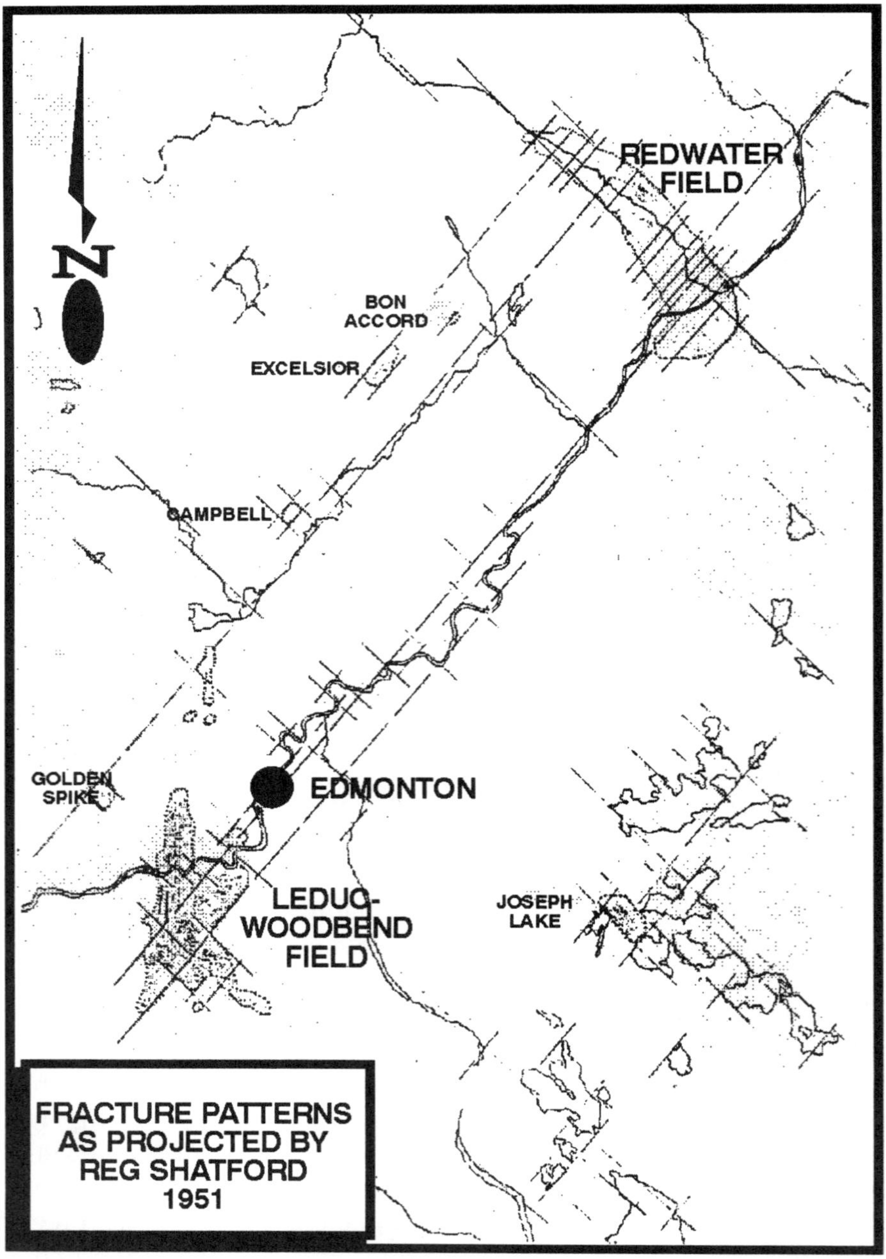

Map of the Leduc-Redwater area showing fracture patterns (R. A. Shatford, author).

Courtesy of Imperial Oil Ltd.

use of the word "stromatolites" to cover that vague never-never land between animal and vegetable life.

Andy recalls his professor in 1947 telling him about the Leduc discovery. The following summer, he visited the Winnipegosis reefs in northern Manitoba and noted that they appeared to be algal. He entered a cave and noted on the wall - in the manner of ancient runes - Ted Link's name with the date 1939 beside it. How much did Link really know about reefs prior to Leduc?

Frank L. Fournier was languishing[8] in the Toronto "Kremlin" when he was assigned to look at Redwater (probably in between his visits to the local "bucket shop" - always playing the market). Frank was well acquainted with the financial world, and it was this connection that brought him back to Calgary, into the offices of Max Bell's Calvan Consolidated.

Frank is acknowledged for an essay in which he compares Eniwetok atoll and lagoon in the South Seas with Redwater:

> ...Redwater reef is approximately 192 square miles in area with a thickness of 1200 feet, giving the bioherm a volume of 48 cubic miles. The reef was built on the Cooking Lake member. Although the corals grew to higher elevations on the outer rim, the overall reef edge attained roughly the same height around its perimeter. The characteristics very strongly suggest we are dealing with an atoll. Since deposition, this part of the Alberta basin has received a southwest tilt of 7 ft. per thousand feet. The strike is N 25°W and the trend of the east edge of the reef is N 43°W. The tilting of the reef caused the oil already pooled there to migrate into the northeastern edge of the atoll to form the Redwater oil field.

> ...The reef lay from one fathom, or six feet, to as much as 200 fathoms below water level. Living reef probably existed down to 20 fathoms or as much as 40 fathoms. At the time of extinction of reef growth, the greater part was living coral - except at the backwaters, where it was tight, fine-grained limestone ... There are upwards of 2000 reef knolls within the lagoon, some of them one mile in width. At Redwater, there could be 1500 such knolls. These features normally grow up to the surface or within ten fathoms of the surface...

Quoting Darwin, Frank says, "In some atolls, all the coral knolls in the lagoon come to the surface of low water. But most frequently, they are quite irregular - some with perpendicular, some with sloping sides, some rising to the surface, and others lying at all intermediate

8 See page 282 "Corridors of Time" for a poetic account of Frank's departure from Imperial.

depths from the bottom upward." Frank asks the question, "Can we find such small features with the seismograph or core drill?" Would the drape of compaction show up on features, say, 150-200 ft. high and half a mile in diameter surrounded by shale? Fournier was, no doubt, anticipating developing plays using his knowledge and Imperial's data, but it seems he confined his activities to partnering with Pacific Pete on crown reserve parcel purchases after he joined Calvan Consolidated.

In terms of more practical contributions, the following geologists each had their part to play:

H. R. Morris was a free spirit. Never mind if his wildcats were busts, he made sure that the drilling was carried out to his demanding standards. Sitting on wildcats was his true avocation. When Redwater hit, Rod's whole life changed. He soon learned that the Jesus Department, once a remote annoyance, was now an overpowering force demanding that he become an administrator. Imperial wanted to drill offsets to evaluate the field.

True to style, he was not too proud to rent the Royer chicken coop for his geological office. After all, what else was available? The author, then with Home Oil, recalls entering the white-washed henhouse and Rod scrunched up at a small desk with his microscope. Kerr was visiting to assess the newly purchased Anglo-Home crown reserve half section. Morris was only too glad to help anyone who visited him and, although run off his feet, always took time out to talk with farmers.

John Downing is especially appreciative of Rod. "He saved my life." As a very junior wellsite geologist John was sent up "clueless - with only a microscope" by Art Nauss of Pacific Pete to look after their newly acquired crown reserve acreage. He recalls Rod explaining the geology to him in a down-to-earth, practical fashion. (No wonder Rod's well-sitting courses of the early 80s were resounding successes.)

Rod received his BSc from UBC in 1943 and had a job waiting for him with Imperial Oil. He and his bride, Jean, travelled to Saskatchewan. Rod's first assignment was at Maple Creek, under Fred Killer and Glenn Fox doing hand auger drilling. Jean fondly recalls their set-up: "Our own tent ... my furniture was orange boxes, and I had a Coleman camp stove with an oven on it. We cooked meals for the hands."

Following his Saskatchewan experiences, Rod sat a deep test drilled by Imperial Oil at Crossfield on a farm-out from Cloakey and Mewburn, two Alberta promoters, and involving Anglo-Canadian.

The location: 9-11-28-2 W5 was spudded in September 1945 and finally completed in April 1946. Despite oil and gas shows at the top of the Mississippian, the well was abandoned. This is where Rod hardened his already jaundiced opinion of engineers and managers. Five years later, just a short distance away, Canadian Superior completed the "discovery" in the same zone.

Rod and Jean then moved to Tofield and Minburn, where Imperial were drilling gas wells to supply the proposed synthetic gasoline plant . But Rod's tenure as District Geologist at Redwater was predictably shortlived. Administrative duties were getting to him, and it did not take much urging by C. B. (Boss) Barlow, Western Leaseholds' ramrod, to sit on their wells. He was then transferred to the Calgary office in the Michael Building, but this did not last long, either. The next move was 2020 Bowness Road in Calgary. It was from this office nearby that, later on, he set up his consulting practice.

Rod was associated with several companies: Osler, Hammond & Nanton and a combination of U.S. oil firms. Each of these outfits had its own "Jesus" department, and Rod would say, "None of them have ever been out on a well, and they're telling me what to do." His experiences led him to confrontations on wildcats from the Arctic to Guyana. Rod always stuck to his word by not charging more than he felt was fair despite the complaints of some of his associates about not getting his prices up.

He found his real métier in teaching the art of well-sitting. Hundreds showed up for his courses, each completed with a written exam.[9] His widow, Jean, remembers as many as 35 pupils in one class and Rod returning home, drained. "But he had his determination and his principles. Somebody said to him, 'Why don't you charge more, say $2000 a pupil?' The answer was 'The oil industry has been good to me, and I want to give something back to it'."

During his career as a wellsite teacher, Rod Morris referred to Redwater as an example of changes in oil fill.[10] Certain wells in the southwest part of the Redwater complex are oil-stained but contain only water. Rod, therefore, deduced that the reef had tilted northwestward prior to the Laramide orogeny. Tilt then switched 90° to the southwest instead of the northwest. To what extent this change created or destroyed reservoir capacity can only be surmised.

---

9 Jean McLaws attended one of the courses while with Indian Minerals and has retained an exam paper meticulously marked by Rod.

10 Don Hoyda, one of his many pupils, recalls Rod's dissertation in this regard.

Doug Lavoie began his career in the Oil Patch one afternoon, in 1951, on the campus of the University of Manitoba. He was called in to an office to take a long distance telephone call from Frank Fournier, who offered him a job. Lavoie already had offers from three mining companies but thought that Imperial's $215 a month was pretty good, and he accepted.

Upon arriving in Calgary, he received a two-week indoctrination. He remarked at one juncture when he looked at well cuttings through the microscope, "That's just a bunch of mud, isn't it?" He was then sent to Redwater to report to Reg Shatford (by then, District Geologist), who put him to work on writing a report entitled "Fracturing in the Redwater Reef." Lavoie concluded that fracturing was localized at the main terrace edge with the talus slope being cohesive. He recalls Reg as being very theoretical and a "what if" personality, a Rosicrucian who believed that several people could receive messages at exactly the same time, and so plagiarism was not a concern for him.

Lavoie's cross-section, with footnotes, is very valuable in that it shows the fractures in cores, as well as a general outline of the different

Doug Lavoie in the author's office.

Furnished by Author

types of rock. Working east, it becomes dense vuggy limestone along with stylolites. This, in turn, leads into dense vuggy reef growth in place. Still farther east, the reef is organic/fragmental with a granular matrix.

Lavoie has vivid memories of Vern Hunter, for whom he had high regard. He remembers him as being friendly with everybody, yet providing strong leadership. During his six years with Imperial, Lavoie recollected: "Nobody trained like Imperial Oil." Basin and regional studies were part of his educational program. Doug looks on Imperial as "my alma mater". Yet, "It wasn't really very intellectual then ... I had a lot of good times ... I really enjoyed the people - got along with most of them."

Douglas J. Carsted, present-day Imperial geologist, has also studied Redwater thoroughly in an effort to get the last possible barrel. He points out that the course of the Redwater River follows the division between the lagoonal and back-reef facies. The river cuts around highs as it progresses southeast, and the water course on the south side of the North Saskatchewan also follows the south boundary of the productive reef.

Carsted displaying a core from the D-3 in his Calgary office.
Furnished by Author

Carsted provides a final comment: "Over the years, numerous geologists and geophysicists have worked on the Redwater field. The most recent geological studies have been conducted by a number of Imperial Oil staff geologists; namely, Jack Wendte (now at GSC),

George O. Callow (retired), and Hardy Turnbull (retired). Their final company proprietary sequence stratigraphic model, developed in the mid 80s, is still in use today at Imperial's office. This could be said to be the LAST KICK AT THE CAT!

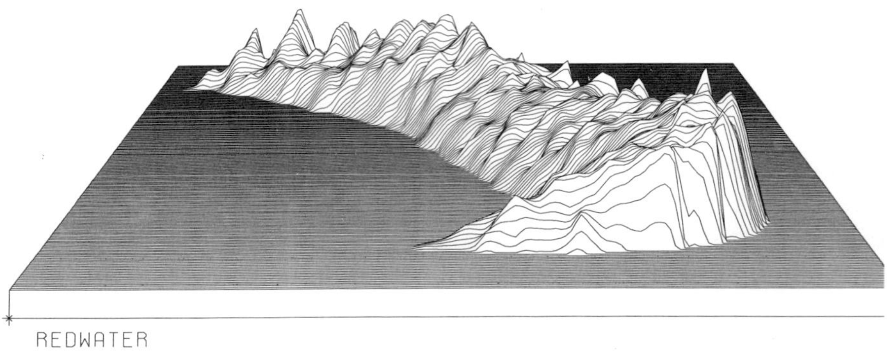

Redwater Reef, looking north-east. Computer generated perspective relief map on top of D-3 porosity. Maximum buildup from original water table 195 feet. East-west true dimension 14 miles, north-south dimension (foreshortened) 15.5 miles, vertical exaggeration 20:1.

Courtesy of Douglas J. Carsted, Imperial Oil Resources Limited.

5 FOR THE ENGINEER

Did the failure to achieve unitization really affect recovery?

This chapter acknowledges the many players involved in the engineering aspects of this giant field. Volumes could be written in this regard, but it is beyond the terms of reference of this book to do more than touch on a few within the key elements:

- Determination of reservoir size, parameters, and capacity
- Attempts at unitization of the field
- Reservoir water disposal systems
- Production engineering
- Movement of the crude to market via truck, railroad, and pipeline
- Town planning.
- Imperial's gas plant construction and development
- Abandonment and clean-up.

In March 1949, Imperial's ball game in Calgary and Toronto changed. Walker Taylor, who had operated the Western Canadian division (not without Mike Haider's behind-the-scenes control), was moved to Toronto to act as an assistant to Haider because of the increasing amount of work underway. Shortly after that, R. B. Curran (Rapid Robert) was brought up from the States to take Walker's place. The Daily Oil Bulletin announced his arrival on April 13th, pointing out that he was a 1933 BSc graduate in Geology and had been a V.P. and director of Carter. Vern Hunter, field superintendent, had some trenchant comments about Curran which the author recorded in a 1983 interview:

> ...Haider (vice president in charge of Western Canada) was responsible for bringing Curran to Canada. Ray Walters was the exploration manager and a hell of a good, honest guy. The only mistake that I ever knew of Haider making was in bringing Curran up here. He was a well-known crook in the States. Everybody who worked for Carter knew that Curran was a crook ... Carter wanted to get rid of him. But why Haider, who had come up through that organization - pretty hard-nosed - didn't know ... He just turned everything upside down...
>
> We had a lot of production battery building operations going on ... government regulations called for fire walls around these tank farms, so that if there was a spill the oil didn't run away. So, naturally, we were building fire walls around the tanks ... (Curran) came up one day with Tip Moroney and asked, 'What are those guys building over there?' Moroney said, 'That's a tank battery, and we're putting fire walls around it.' Curran said, 'We don't need any fire walls.' Moroney:

---

1 Those who wish to delve into more detail may refer to the Bibliography.

> 'Probably we won't need them, but we have to have them because it's regulations.' Curran: 'Well, we're not going to build them.' Tip: 'Are you ordering me?' Curran: 'Yes. Shut it down.' So Tip came over to me and said, 'Curran says, shut down the cats - we're not building those fire walls.' I said, 'You know darn well we can't do that.' He said, 'Do you want me to put it in writing?' I said, 'No, you don't have to put it in writing. If you say, shut them down, we shut them down.' So, I got Leigh Constable, the Production Superintendent, to shut the cats down. No more fire walls...
>
> Well, next morning I got the word - I think it was Rod Edgecombe, local engineer with the Conservation Board. 'What's this about you shutting down the cats and not building these fire walls?' I said, 'Yeah, we shut them down last night.' He said, 'Why?' I said, 'Because I was told to.' He said, 'Who told you?' I said, 'Tip Moroney told me. He's my boss. But somebody else told Tip.' Anyway, he took off, and I don't think it was a half hour later I got a phone call from Tip saying to start the cats up again...

Hunter also recalls Curran promising jobs to his U.S. drilling contractor friends. Moroney would be approached by Curran's buddies (probably at Coutts) wondering where to move their rigs. Apparently, Moroney did his best to put a stop to this, but it was difficult. "It was the same with Ray Walters. You know, Ray didn't want these (U.S.) seismic crews, but what do you do? Your boss tells you to put them to work - and you're hiring rigs and seismic crews - what do you do?" Also:

> ...Bob Curran used to give farm-outs to companies and then immediately phone up the broker and buy stock. He would say Ray Walters had agreed to the farm-out, and Ray had never even heard about it. Well, George Shultis was a pretty darn good electrician, as well as being a good seismic man, so he bugged Curran's desk and the phone so he could record everything that was said. But Ray Walters couldn't stand that very long. He just quit. He went to South America, and I think he died within a very few months afterwards. I never did hear exactly what happened...

Jack Harvie, engineer, confirms Curran's antics: "Ray Walters told me that he kept a tape recording of all his conversations with Curran - he didn't trust him. 'Rapid Robert' Curran was always trying to make farm-outs for his friends."

George Bannantine was appointed field superintendent of the Redwater Area. He had 24 years of experience with Esso, the most recent in Rumania, where he was imprisoned by the communists, tortured, and finally escaped. Leigh Constable was sent up to be his assistant and became the field superintendent when Bannantine moved back to Calgary. Gibby Gibson thought it a little unusual that Bannantine was chosen as the first field superintendent. Why an international expert? Why not a Canadian - such as Vern Hunter, Leduc's first boss? (Hunter eventually succeeded Constable.) However, according to Gibby, he soon hit it off with George:

> ...Bannantine's prior experience had been with the Jersey, or Exxon, organization and more on the worldwide scene ... His background was certainly drilling and production but, as he told me shortly after he arrived in Redwater, "I don't know how much help I'm going to be on the day-to-day operations. I've spent most of the last ten years of my career in the court house" (laughter). His dealings were government level rather than operational. He was sure a good man to have around. I remember on one occasion we were reviewing our cost, and with this comes manpower and overtime ... He says, "All that overtime tells us one of two things. Either we haven't got enough people or we're running an inefficient shop - one or the other. We shouldn't be working that much overtime"...

An evening at "The Troc" in Edmonton. L to R, Jack Gallagher, Emily Naldrett, Maurice Paulson, Ev Paulson, Gibby Gibson, Marg Gibson, Hugh Naldrett, Cathy Gibson (Gibby's sister).

Furnished by W. J. Gibson

Ray Walters was edging in, becoming assistant exploration manager under Jack Webb, to help relieve some of the pressure. The pressure increased on Webb to the point where he finally he packed it in with Imperial on November 25, 1949. Ray Walters took over, pushing Ernie Shaw - another contender for Webb's job - to one side.

Maurice Paulson was the first district engineer, Willis (Gibby) Gibson was district petroleum engineer, and Jack Harvie was district civil engineer. Bill Cousins was head of production, Bernard Jeffery supervised accounting, and Stan Pooler ran the warehouse. Harry Robertson was in charge of construction and transportation.

Jack Harvie describes the situation when he became involved with Redwater:

> ...Dave Gustafson was the chief engineer ... I'd pretty well wound up my assignment in Calgary, so I said to Gus that I'd like to go to the field. He just sent me up to Redwater. Maurice Paulson was in Redwater then as head engineer. And Gibby was a Production and Drilling man ... and I took over the construction, flow lines, roads, well site preparation, and all that. In Calgary, Vern Taylor was Walker Taylor's (no relation) assistant, and then Charlie Visser was the drilling supe ... Ivone Burn was in charge of Land. Henry Lamarchand Stevens-Guille ... from Turner Valley. He was sort of in charge of the engineering. I reported to "Steve". And then George Fisher was in charge of Purchasing. And then all the Exploration guys under Jack Webb were in the Tecumseh Building, along with the draftsmen and so on...

Imperial also formed a Joint Industrial Council with the intent that its members be the voice of the workers. The Council met monthly, in Calgary, with transportation and accommodation paid for, dinner and drinks after the meeting, as well as a full day's pay. It was a clever ruse to prevent the formation of a union and, surprisingly, it succeeded. Everyone seemed to be happy with the way it operated.

Amidst all the politicking, workers were busy trying to determine Redwater's size, parameters and capacity. The discovery well had shown under-saturated crude with a low GOR and little or no drawdown, indicating a huge reservoir. Judging from the final drillstem test, it appeared that there was an effective water drive.

Frank Black, one of the first Canadian Schlumberger engineers, is the proud owner of an important memento. Mounted in a frame is Run #2 at Redwater No. 1, the original copy of the log he ran when total depth had been reached. Frank recalls running in his sonde to

Joint Industrial Council, 1949: G. G. Cochran, Roughneck - later Safety Man; J. W. (Wilf) Bowe, Roughneck; C. W. (Charlie) Beck, Production; C. A. (Carl) Laudan, Transportation; Joe Jackson, Toolpush; W. O. (Bill)(Curly) Blinn, Toolpush; W. O. (Bill) Twaits, on assignment in Western Canada, learning the ropes in preparation for his presidency of Imperial Oil; T. C. (Tubby) Peacocke, Drilling Supervisor; J. H. (Harry) Morris, Toolpush; F. C. (Fin) Lineham, Toolpush on Redwater discovery; J. E. (Johnny) Bruce, Cathead Man on Joe Jackson's rig; G A. (Gerry) Gingras, Driller; A. J. (Jumbo) Bennett, Driller; J. W. (Joe) Henry, Driller; C. R. (Carl) Olson, Derrick Man; C. (Charlie)(The Dutchman) Visser, Drilling Superintendent; G. H. (George) Fyfe, Driller; B. J. (Tip) Moroney, Production Manager; L. A. H. (Hec) Marlowe, Driller; W. D. C. (Don) Mackenzie, Chairman of the Joint Industrial Council and Western Division Manager.

Furnished by Audrey Bowe.

the bottom of the hole and turning on his equipment; there was no galvanometer response. He thought there was a bad connection in his circuitry, because he was sure he was on bottom. He checked the depth with the driller and started to pull up when, bang, as soon as he hit the oil-water interface, the galvanometer flew over, showing high resistivity. Frank had logged below the oil-water interface in the water-saturated section of the limestone. The driller had put on an extra single, not knowing he had drilled into water. This is the first time, ever, that Schlumberger had logged water and realized that they could now read fluid content in carbonates. Measuring porosity directly would be a challenge for the researchers.

Frank recalls the tremendous boom in drilling. The pressure was not without its victims. On one occasion, the door of the recording truck opened and a body fell out. The man was so sleepy, he couldn't stay awake - just fell out onto the road. Frank Black, himself, was run into the ground - all 120 lbs of him.

H. N. (Neal) Collins graduated from U of A in the spring of 1947. (Don Harvie, of Western Minerals, was a frat brother.) Close to graduation, on his way home to Lethbridge, he stopped off at the Calgary CPR station to be interviewed by Charlie Bailey and Frank Black of Schlumberger. He also received an offer from H. O. Medsger, Canadian manager of Heiland in Edmonton, but decided in favour of the former.

H. N. Collins, P. Eng.
Furnished by Neal Collins

The boom in Redwater was unbelievable. Collins gauges the height of activity as being in the spring of 1949 - everybody running out of their pocket. A little later, Schlumberger installed a quonset hut. When Neal went out to log a well, he would leave a sign on his hotel door saying which well it was and when he'd be back. His driver, Harry Wright, could usually be found at Pete Crow's poker game.

Between June and September 1949, Neal reckons he logged over 200 wells (58 in one month). His base pay was $200 per month with a

$2.40 daily living allowance. Engineers received a bonus of $6 per job and truck drivers $3.20 each. There was no such thing as overtime. The late Jack Minchin referred to Collins as "money-bags" when he pulled down $700 in one month. A notable exception for which Collins was poorly remunerated was his logging of the hotel water well, receiving a case of beer for his expertise.

According to Jack Nesbitt, Imperial geologist, Lane Wells laid claim to the ability to measure porosity with their neutron logs. This was said to have sent Schlumberger into a tizzy, so they brought in experts from Paris and Houston in an attempt to duplicate Lane Wells' work. Collins discounts this story. It is true that Lane Wells were taking a lot of business away from Schlumberger. The latter company had only the MicroLog and E-log. There was no Laterolog at that time. Halliburton's caliper log was unsuccessful in competing with the microlog.

M. P. Tixier and **R. L. Forsythe**, two senior Schlumberger engineers, wrote a paper in 1951 (which appeared in the CIM Bulletin) describing the MicroLog and the LateroLog. Fig. 5 of their paper shows the effect of porosity and creation of mud cake to give a curve separation, but there is no quantitative porosity measurement. The LateroLog purports to record deep penetration, thus achieving true resistivity by eliminating the effect of the mud in the bore hole. The paper tries to explain: "Continuous efforts are being made toward an improved interpretation of the conventional logs and toward the development of new methods which may permit a more accurate and detailed investigation of reservoir characteristics." They still had not come up with a sonde to compete with the neutron log of those days.

Directly following the paper are comments by Jack Browning, then of Bear Oil, and Robert Pot, reservoir engineer with Imperial Oil. Browning emphasizes correlation over long distances using certain key markers such as the Viking. Pot states, "It would, therefore, be of primary importance to establish a value for as many factors as possible of the formations actually encountered in Canadian fields."

In terms of reservoir engineering, worthy of note are J. E. (Ted) Baugh's paper of December 1950, listing reservoir parameters[2] and F. L. Fournier's tabulation from an Imperial Oil report.[3]

---

2 Appendix X.
3 Appendix XI.

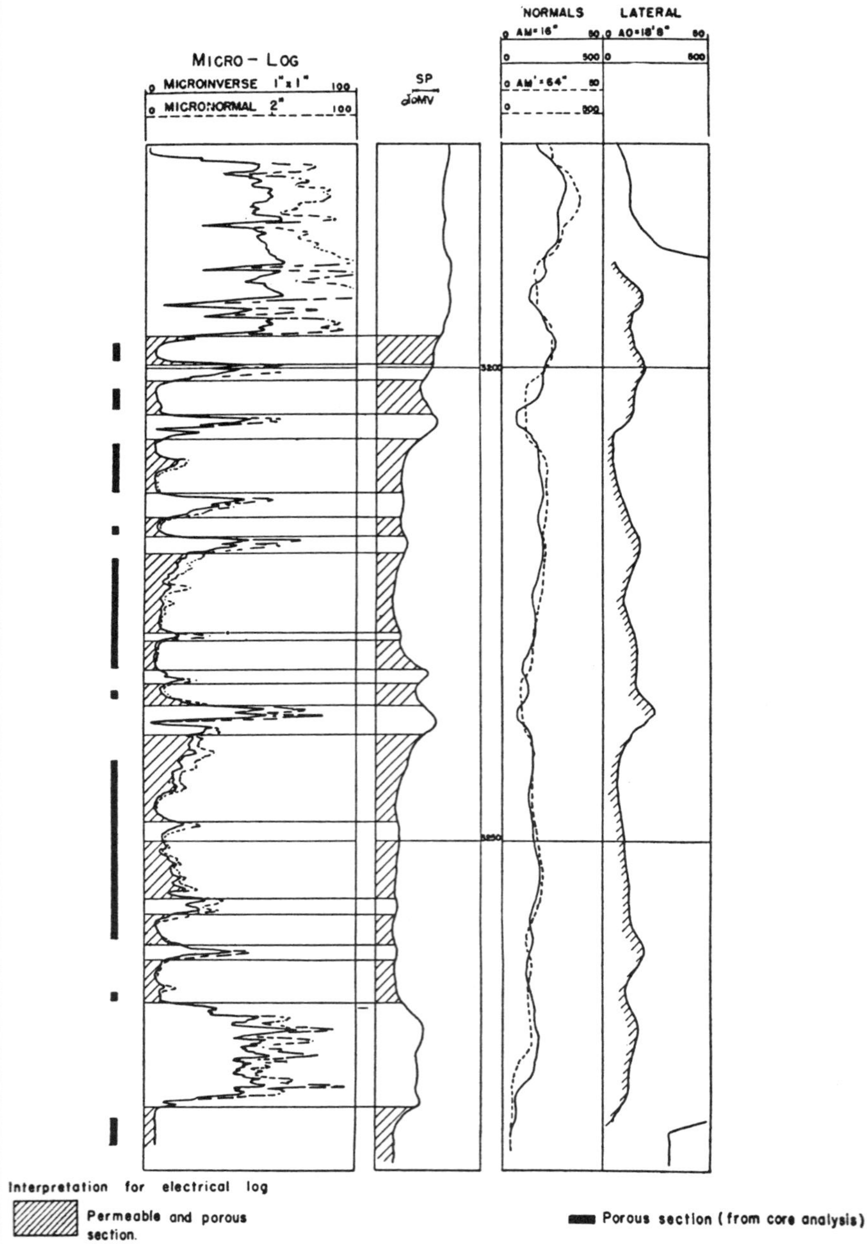

Figure 5.—Example of MicroLog, Redwater field.

(Amplified depth scale)

FIG. 5 from Tixier and Forsythe's paper illustrating microlog.

Irene Haskett, retired engineer in Dallas, describes Redwater, "My favourite. Just a pity that it did not receive water injection ... one of the most mistreated reservoirs, so ideal for pressure maintenance. This could have been accomplished before reservoir pressure dropped below the bubble point." She laughingly stated that all one would have to do is drill holes along the Redwater Creek river bottom, enter the reef, and let the river water flow down and do its pressure maintenance. Irene Haskett was a reservoir engineer from Carter Research assigned to Calgary in 1949. "Gibby" Gibson recalls picking her up at Bob Kashower's Edmonton Airport Motel (quonset huts) to drive her to the field on her first day. Irene remembers discussing geology with Reg Shatford, who did most of the lithological work, including reef tops.

Irene Haskett with Core Labs, circa 1955.
Furnished by I. Haskett

She took six months' leave of absence from Imperial in order to travel. Prior to leaving (she said she was bored) Irene turned out *the* definitive engineering report[4] on the reservoir. It was subsequently published in the CIM Bulletin, April 1951.[5] In it she cites the research of John R. McKay, petroleum engineer with British-American, and

4 Roly Horsfield, Imperial engineer, was the ghost-writer for Irene's paper.
5 Because of its excellence, it is reprinted in its entirety as Appendix XII.

Jack Henley, Lane Wells, who collaborated on a paper establishing a relationship between porosities as derived from cores and "pen travel." Their work, published in the July 1950 issue of COGI,[6] presented two important concepts:

- Neutron curve response is a function of formation fluid content.
- Formation fluid content is a function of porosity.

Haskett's report pioneered in core analysis related to neutron logs - one sample per foot of cores got a good correlation. Values were smoothed out by using two-foot weighted averages. She clearly identifies the reservoir: 900 ft thick, 200,000 acres - 37,500 of which are productive. The back reef runs into Bon Accord. In contrast, the Morinville reef is of the same age but dolomitic and believed to be entirely separate. The top impermeable layer has dense limestone streaks scattered throughout. The highest reservoir capacity is found in the sand-like intergranular material plus vugs along the edges of the reef. It concludes, "...The relative amounts of void space and the vuggy material which have low porosity occupy the largest portion of the reservoir. The smaller volume of high porosity intergranular material is probably the most important factor. Much of the matrix porosity is automatically eliminated due to low permeability..."

When Irene returned from her leave, she received a call from Pacific Pete to come over and work for the producibility faction - their first engineer. "It was a fun outfit. I never knew what they were up to."

Unitization negotiations of 1951-54 were hinged on rock types and the issue of porosity in the hope they would resolve tract factors (the amount of recoverable oil under a given 160-acre tract). Nearly 1000 well logs, samples and cores, not to mention the newfangled neutron log, were worked through to arrive at reservoir capacity for each well core, each a tiny piece in the jigsaw puzzle. The first formal step towards unitization was the creation of the Redwater Operators Committee.

The **Redwater Operators Committee** with top representatives: Gordon Webster (Home Oil), W. J. "Gibby" Gibson (Imperial Oil), Gordon Connell (Royalite), R. H. "Rich" Swann (Western Minerals), to name but a few. This group, in turn, appointed subcommittees,

---

6 Canadian Oil and Gas Industries.

Christmas 1950 office party: L to R - Bernard Jeffery, accounting; Bill Smith, land man; Ralph Flanders, production; Rod McDaniel, engineering; Ed Cutlin, engineering; Ron Shirra, engineering; Julius Abzinger, engineering; Lauder Nowers, personnel manager; Ben Hochhausen, engineering; Eric Connor, engineering; George Bannantine, field superintendent; Harry Robertson, transportation and construction

L to R - Harry Robertson, Reg Shatford, geologist; Bill Cousins, production superintendent; Burt McDonald, production foreman; Bud Kelly, drilling superintendent and Millard van der Grift, production foreman (leaning over); Maurice Paulson, district engineer, Norm Wright, production foreman (with doily on his head); Jack Turner, engineering; Bob Frances, engineering; Jack Nesbitt, geologist (in background), Jim Henderson, engineering; Doug Amos, production (leaning over in front of Jim); Elliott Van Deusen, production; Stu Bernard, accounting; Ken Oakley, engineering; Si Cormack, production; Jack McRae, Willis Gibson (holding teacup); Stan Pooler, warehouse and Hugh MacKenzie, production (crouched in front).

Furnished by W. J. Gibson

the main one being the Redwater Working Committee, responsible for studying reservoir characteristics:

R. D. Knowles, Chairman	Imperial Oil
Irene Haskett	Imperial Oil
G. Haugrud	Royalite
J. Maughan	British American
R. Shatford	Imperial Oil
R. Woods	Gulf Oil
J. Strand (Alternate)	Gulf Oil
R. Gilmore (Alternate)	British-American.

The Operators Committee was presented with three positions:

(1) Imperial Oil appeared to be neutral simply because they had so many, and such a variety of, reservoir types.
(2) The "have-nots", which included British-American, Gulf, Hudson's Bay, and Pacific, were more for a flat allowable and did not want to use porosity as a parameter.
(3) Royalite and its neighbours advocated porosity to determine allowables because they had some of the best reserves.

Dave Gustafson of Imperial Oil was Chairman of the Redwater Operators Committee and Irene Haskett was his assistant. They were augmented by Carter Research experts from Tulsa, who were adamant as to the need to unitize and waterflood. To quote Rich Swann, Western Minerals' senior engineer: "They had the dope to prove it." Carter's position made everyone suspicious of Imperial's motives. R. D. (Doug) Knowles, former senior engineer with Imperial Oil, and at one time chairman of the Redwater Operating Committee, felt there was an intent to unitize. The reservoir engineers ran tests in the Carter lab at Tulsa using The Pod[7] and they were trying to duplicate the water drive. Their results indicated there was insufficient water drive in the reservoir.

Reg Shatford, Imperial geologist, is said to have predicted that water would be coming in from the ends of the field, not from the sides. Further tests, still using The Pod, indicated there was, after all, sufficient water drive. The line pipe purchased by Vernon Taylor was stacked and ultimately used for other purposes. Art Bain, Imperial engineer, recalls the good permeability and bottom-water drive: "To most people, it seemed equitable to unitize on the basis of oil-in-

7 Abbreviation of Podbielniak, analyzer, which fractionates liquid hydrocarbons.

Doug Knowles (left) with unidentified man at Norman Wells.
Furnished by R. D. Knowles

place or reserves (assuming the same recovery factor for the whole field, which wasn't true."

Warren Davis, one of the top Gulf engineers from Tulsa was closely involved in the proceedings. He wanted the formula to give each operator the same rights as natural depletion; that is, a flat well allowable. The reservoir was under-saturated and a fast pressure decline could mean a recovery as low as 20%. Jim Strand, former Gulf engineer, recalls the Working Committee bringing together all the Redwater parameters and putting them on maps. Meetings were held in the Michael Building (3rd Street S.W., Calgary) courtesy of Eric Harvie of Western Minerals.

Mr. Roper, brought in from the States by Western Leaseholds, expressed great concern that insufficient homework had been done by Imperial, saying there was no need to go to waterflood, let alone unitize. In effect, he said "Let's go our own way." Everyone was afraid to have production controlled by a unit; that is, the rule of equity prevailing over the rule of capture. Rich Swann claims the operators tried to get the Conservation Board to declare that capture was not equitable.

Gordon Webster, who represented Anglo and Home at the unitization meetings, remembers Pacific Petroleums' strategy - not

L to R: Willis (Gibby) Gibson, skip, Maurice Paulson, second, George Wright, third, and Art Bain (kneeling), lead. "We won prizes. (a small mantel radio) for having been good enough to beat Campbell Aird when Gibby made a magnificent draw on the last rock," said Art.. "Some of us then spent the night in sleeping bags at Rod and Marilyn McDaniels' home."
Furnished by A. Bain

wanting to talk about reserves when it came to tract factors. Even though their wells were over on the west edge and poorer, they were all making their allowables and wanted to base their equity on that. If they had been forced to go the reserves route, they would have ended up with only a fraction; therefore, Pacific's position was regarded as an obstacle.

Royalite's Gunnar Haugrud recalls arguing with his team over the top of the porosity to within a half foot. The west side of the field had bands of porosity interlayered with dense zones which yielded more oil than wells to the east, which watered out first. Gordon Connell's stand, that Royalite should receive higher allowables for their wells on the east side, received little or no support.

The Working Committee produced a report, in April 1953, of a well-by-well analysis detailing the reservoir parameters. This was referred to as the Green Hornet Report because of its distinctive green cover and the popularity of a radio program at that time[8].

As a result of this research, it was predicted that, with the declining field pressure, the relatively inefficient combined water and internal gas drive could replace the present efficient water drive. To

8 See Appendix XIII for sample pages.

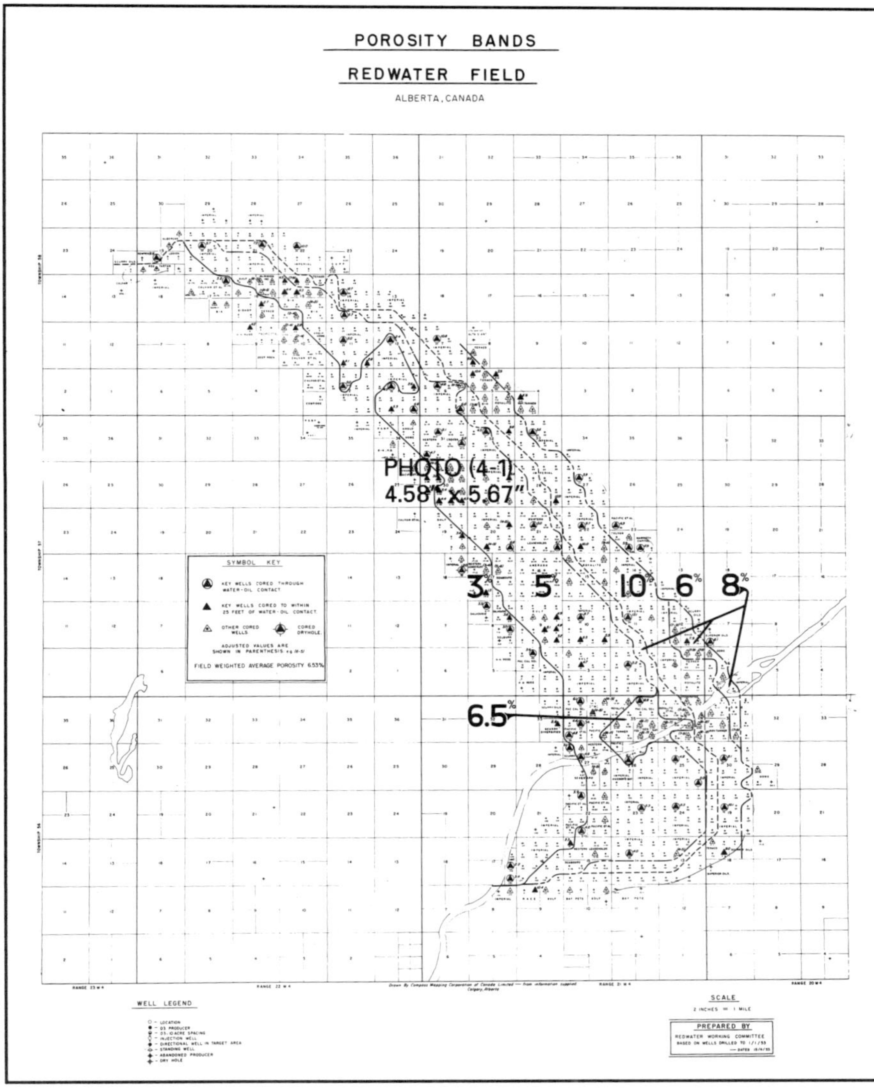

Plat prepared by Redwater Working Committee showing per cent (as of Jan 1, 1953) porosities.

Courtesy of Imperial Oil Ltd.

offset this expected production decline, a pressure maintenance program was recommended for the early part of 1954. Along with these technical matters, there were serious legal issues to be addressed.

The **Redwater Unit Legal Subcommittee (RULS)** was formed by the Steering Committee of the Redwater Pool Producers

Committee (RPPC) on March 13, 1953. They were to "consider and advise on the legal aspects of unitization of the Redwater field and draft the necessary statutory amendments to accomplish such a task." On June 15th, RULS recommended that the following steps be taken:

- Secure legislation to legalize the proposed agreement;
- Furnish solutions to taxation and accounting problems;
- Investigate land titles which would be subject to the proposed unit;
- Seek consent from the Government for the taking of water, for reservoir flooding purposes, from the North Saskatchewan River.

On November 20, RULS produced recommendations calling for amendment to five statutes:

1. The Companies Act provided that no company association or partnership of more than 20 persons could be formed unless it was registered as a company under the Act. It was obvious that more than 20 individuals would be participating in the unit. The idea was to amend the Act so as to empower the Minister to exempt the unit.
2. Add a new Part IX to the Mines & Minerals Act authorizing the Minister to enter into an agreement for a unit operation. Further, the operational requirements of each Crown Lease within the unit area would be met in such a way that the product allocated by the agreement to a tract would be deemed to have been produced from that tract.
3. The Mineral Taxation Act required amendment so that production allocated to a freehold parcel would be deemed to be production for taxation purposes.
4. Section 95(1) of the Oil and Gas Resources Act of June 1, 1950 put the onus on the Petroleum & Natural Gas Conservation Board to promote and assist the pooling and unitization of oil and gas interests. Subsection (2) called for the Board's approval for any scheme, while (3) required the owners of oil and gas interests to obtain a permit to allow for development and production. It should be noted that the Board had no jurisdiction over a unit operator and an amendment was needed to allow the Board to deal directly with that entity rather than all the diverse owners and operators. This was accomplished by adding a new subsection (4) to Section 95.

5. An amendment to the Land Titles Act regarding the filing of unit agreements and the necessary endorsements on the Certificate of Title so affected to ensure that future purchasers would be bound by the existing agreement.

In December 1953, A. E. Keller, assistant manager with Texaco and chairman of the RPPC, wrote to Premier Manning regarding these recommendations. Also, in January 1954, C. D. Williams, Chairman of RULS, met with H. H. Somerville, Deputy Minister of Mines and Minerals. Williams was referred to Deputy Attorney General Harry Wilson in regard to the Land Titles Act. The five statutes were duly amended and came into force April 8, 1954.

One can say with authority that C. D. Williams, together with Hubert Somerville, set up the legal machinery for all future voluntary unit proposals. The earliest attempts at unitization are reflected, as Hubert Somerville points out, in a 1941 Act.[9] Designed to bring the Crown and freehold (as owners, not lessees) together for development on a unit basis, this statute was used in only four cases, because there was no procedure whereby royalty/working interests could participate. It was, therefore, unworkable for Redwater.

By the end of 1953, the Redwater Operators Committee had been unsuccessful in establishing a formula for unitization of the field. As a result, it was decided to engage consultants. Messrs Keplinger and Wanamaker represented Gulf, B-A, and Pacific. Kaveler acted for the pore volume operators. The three met over a period of several days, hot and heavy, with no resolution. Irene Haskett recalls Keplinger as shallow in his thought processes, demanding pro-Pacific producibility as the key equity factor. Kaveler, on the other hand, claimed porosity was the only factor. Kaveler "would testify to anything" but, when caught off-base, he would counter, "I've learned since." Warren Davis went around the table but was unable to reach a consensus. It was declared an impasse. Everyone went away, pressure stabilized, the ocean came in and there was natural waterflood about six months later.

Paul Greenwood, former Conservation Board engineer, sums up this unitization attempt quite perceptively:

> ...Unitization of the field was mooted in 1951 or '52 by Imperial. Imperial really considered that Redwater was their field - since they had over 500, I think it was about 550, wells out of the 920-odd that were in the

---

9 Repealed April 8, 1954.

Paul Greenwood, circa 1950.
Furnished by P. Greenwood

field.. and all the rest were interlopers. So what was good for Imperial was obviously good for everybody else ... The fact of the matter.. there was really no good reason to unitize because there was no real reservoir enhancement reason. The natural water drive supplied from the Morinville Reef gave an expected 60% recovery factor without doing anything more than returning produced water to the formation. Therefore, the upshot of all the unitization hearings was the formation of the Redwater Water Disposal Company. It really turned out to be a good thing, because in those days the Board gave barrel-for-barrel credit. So, if you reinjected all your produced water to the production formation, you got full credit on your oil allowable. I imagine that meant more to the operators than anything else...

The Conservation Board had little involvement with unitization - it was just as well, because the operators never did agree on a formula and the whole issue simply "fizzled out". However, salt water disposal was a problem that the Board had to address. Paul Greenwood outlines the situation:

...Salt water production was always a major problem in Redwater, right from Day One, especially on the east flank. Large quantities of water were being produced as early as 1950 and '51. Pit disposal, of course, was the standard method. And because of that type of disposal, the wells were subject to a water-oil ratio penalty factor on the crude

> production. Imperial Oil had a huge salt water disposal pit out in the sand hills east of Redwater on the east side of the Redwater River.
>
> Salt water was trucked in seven days a week, 24 hours a day, and yet there was only a very small amount of water in the pit along with the usual oil and scale scum. In 1951 the Board became concerned about what was happening to this salt water, and an engineer from Imperial named Jim Henderson and I conducted a survey. We tested all the (shallow) homestead water wells, most of which were hand-dug, and sampled the Redwater River from near the pit right down to where the river entered the North Saskatchewan. The only interesting fact was that all the homestead water wells were contaminated with a high amount of nitrates. We never found any indication of high salt water concentrations or abnormal concentrations in the Redwater River anyplace...

There appeared to be very few environmental concerns, but the disposal of the ever-increasing volumes of water had to be resolved.

The Redwater Water Disposal Company (RWDC) was formed to handle this. Frank Manyluk, engineer and Conservation Board member, describes its formation and resulting success:

> ...As the field developed ... it was decided that it would be more advantageous to have a SWD company who would build lines up and down the field, connecting to most, or all the batteries. And it developed into ... one of the biggest SWD companies in the world! A very efficient company, well run ... I'm not a reservoir engineer, but I think, in the end, wound up in the recovery of more oil ... by slowing the decline in the reservoir pressure ... Ben Hochhausen was one of the real reasons for its success...

E.F.D. (Ben) Hochhausen was born in 1924 on a farm south of Provost, one of ten children. While attending high school in Hanna, one of his teachers suggested that he become an engineer. To Hochhausen that meant "one of those guys that drove a locomotive." However, he later graduated from the University of Alberta with a degree in Civil Engineering.

Gibby Gibson, Imperial engineer, takes pride in putting his finger on this new graduate and bringing him up to Redwater. Ben's first job was to help Jack Harvie with "construction, flow lines, roads, well-site preparation, and all that." Gibby then assigned him to locate and design tank batteries even though, as Ben says, "I had no idea, in oilfield jargon, what a battery was." As water production increased to the point where surface facilities became inadequate, Gibby re-assigned Ben the job of gathering and disposing of the unwanted

Ben.
Provided by E.F.D. Hochhausen

brine. This was to become Hochhausen's pursuit, escalating him into a world-class career.

In 1952 Imperial brought in an engineer from Kansas, Ivan Rice, a consultant who had worked for Gulf Oil and was experienced in designing water disposal systems in Kansas. Ben was loaned to Ivan - a relationship that was to have a profound impact on both parties, as well as Redwater itself. The RWDC was formed in 1952 involving Rice and operators in the Redwater field. Hochhausen switched over to RWDC. He recollects its structure:

> ...800 wells, and each company was given two shares for each well, except Imperial -which would forever hold their shares down to 45%. That entitled them to five directors on the eleven-man Board. The first Chairman of the Board was Vern Taylor, and the members were Tip Moroney, Vern Hunter, and Loughney (Gulf), Powell (Seaboard), and Gordon Connell (Royalite). The meetings always took place in Calgary. Each company would put up $2500/well...

Hochhausen describes the company that he and Rice formed during this period and of which he later became President.

...We formed Rice Engineering & Operating Ltd. with the U.S. people having 50% of the shares of the company and the Canadian people having 50%. I, at that time, had 30% of the 100; the two U.S. people involved had 25% each; Dave Hindley had 5% - he was the fellow whom I stole from Imperial when they first started. And Les Hunter, who was a chartered accountant, had 15%. Les Hunter was Vern Hunter's brother. Interestingly enough, with Harry being in the gas business, Vern in the oil business, and Les in the water business, we had a little joke about it. We operated with a simple agreement for 20 - 25 years, which enabled Rice to take on other projects outside of Redwater field...

The success of this operation was dependent on inter-company co-operation and a great deal of initiative and innovation on the part of Rice Engineering, Hochhausen in particular. Hugh Mackenzie, who did consulting work for Ben, held him in the highest regard: "Ben had a way of looking at things different than any man I ever worked for ... He looked at things in kind of an abstract way. And he came up with some of the greatest ideas going. He was way ahead of his time ... the brains behind the whole thing."

It certainly took brains to keep up with ever-increasing production of fluid. Initially, they were looking at 200 to 500 barrels per well per day, maximum. The advent of the "REDA" (an electric motor submersible pump) changed all that. Eventually, some wells were producing 4000 - 6000 barrels of fluid per well per day. To quote Hochhausen:

...Our production got so great, and we needed so many more wells, that we started to have to put in pumps, and so on. And we decided then that what we should do was utilize some of the pressure, or the power, they had at these various field gate batteries, and we redesigned the complete system as a two-phase gas-water, slug flow gathering system ... There's 48 injection wells out there now, and many of them can take up to 40 and 50 thousand barrels of water a day by gravity. There's no injection pumps. The wells operate on a vacuum, and we utilized that vacuum when we designed the system to real advantage...

An unexpected outcome from this expansion is explained by Hochhausen:

...Oil sales that we made from skimming oil from the water. One year, we sold $7 million worth of oil off the water, which belonged to the operators. But we credited the operation.. We had to reorganize the company completely - so that we had classes of shares for each

Automated well controls.

Injection wellhead in cellar
Furnished by Author

operator, and the amount of dividend they got was dependent on the amount of water they produced. This assumed that every barrel of water had the same amount of oil in it...

There were times when we knew that an operator had an upset but, once it hit our pipe line, it was our oil. And the revenue would go into the Redwater coffers. So we had to reorganize to handle the produced oil. And it wasn't very much: 6 barrels per 10,000 bbl of water...

As with many other oilpatch endeavours, water disposal has its tales of humour and danger. Hochhausen had his share, from woodpeckers to sour gas:

...At the batteries, we had a wooden gas boot - metallic ones would not stand up -20-foot high, and we took the water from the treater, and dumped it into the top of that wooden boot. The water in the boot sought its own level - water elevations sufficient to boot the water down the pipeline to the disposal well - and the gas went off the top ... In the Amelia area, it was rather interesting: The woodpeckers would hear the water gurgling into these wooden gas boots, and they would start pecking holes in them, thinking they heard insects ... We had to take one of them down because there were holes in it ... and 27 dead woodpeckers...

As for the increasingly dangerous H_2S problem:

...When we started producing all of that water, and started taking the water from the satellites, we then had a large amount of gas with the water - up to four cubic feet per barrel ...It was rotten ... 17% H_2S (lethal)! Nobody wanted it. We couldn't gather it ... When you look at a million barrels of water and an average of 2 cu. ft/bbl, that's two million cubic feet ... The Environmental people ... started putting H_2S monitors in the field, and they traced a lot of it back to us...

It was then that we developed our incineration devices. We would take the gas off the top of these concrete tanks and still maintain a few ounces of pressure on the tank. We put in a stainless steel suction fan to maintain a slight vacuum in the tank. And the gas was burned in a refractory-lined vessel at temperatures up to some 800 or 900°C and cracked to SO_2. This incineration was very successful ... it eliminated our environmental problem...

Hochhausen also extended his innovativeness into other areas:

...We developed a pipeline insulating nipple that we can operate up to 1500 psi ... that I've got a patent on ... Esso was very interested and very cooperative ... they opened up their lab ... 'Whatever you want to test, under whatever conditions, we'll do it for you' ... I had thousands of dollars worth of tests run ... Those nipples are being sold right through the States now. They screw them in between the wellhead and the flow line to stop the flow of cathodic electric current. It's simple. I don't know why anybody hadn't thought of it before...

Just a few years ago, Ben turned Rice Engineering over to seven of its employees, with certain instructions and requirements - they've been more than met - they're still doing a good job in Redwater, as did Hochhausen.

Frank Manyluk indicated that 99% of his Conservation Board activities evolved around water disposal, drilling and production, with emphasis on the latter:

...My biggest problem during my two years in Redwater was getting the drilling companies to fill out the tour reports. I always remember, the drilling companies.. used the tour reports as time sheets. And there was nothing on it about depth, surveys - all the things that were supposed to be on it weren't.. It developed into two or three cases - and once they got settled, where we would shut a rig down for not

Frank Manyluk, dressed for the great outdoors.
Furnished by F. Manyluk

keeping proper tour reports - it got around to the others, and they started to keep proper tour reports...

There were unsuspected perils, too:

...With all the activity - the drilling, the flare pits.. little fires would start in the muskeg, and then go undergound, and travel underground for at least a half a mile, and then pop out again, and then go back under again.. And there were numerous fires of that sort right from Simmons.. up to Opal. There were some very serious ones at Opal. It pretty near got some farm houses! The best thing you could do is guess how far the fire had gone and ask the.. company drilling the well to take in a "cat" and dig down and establish a backfire ... or, at least, not let it spread past that point. And, eventually, they burned themselves out...

Gibby Gibson provides an overview of drilling and production at that time:

...The work in Redwater, the development side of it, was pretty straightforward, although we were sure busy. Redwater was a rather easy pool to develop. It was relatively shallow, 3000 ft. plus, no gas cap. Wells were drilled fast. In our heyday, we were drilling and completing a well just about one week per rig. There were some problem areas. In some sections of the field, there was shallow gas

> that gave us trouble even setting surface pipe. We had a few blow on us there, but no major ones...
>
> Some of the edge wells gave you completion problems as you got near the feathered-out edge, and the water was lying underneath. This was particularly true on the eastern edge where there seemed to be some vertical fracturing, and we had trouble keeping the water out as we completed the wells...

Some of the more prominent players were:

Hugh Mackenzie, whose first oil patch job, at age 14, was in 1941 with Purity 99 Refinery, became a battery operator at Leduc before being transferred to Redwater to work on completion. One of Hugh's responsibilities was the maintenance of treaters. He recalls anode tubes lasting only six or seven weeks, on the average, sometimes less:

> ...Some of them would go in as little as two weeks ... before they discovered they could put a coating on these tubes to reduce the amount of corrosion and increase the number of anodes on the body of the treater - so that it was protecting the body more than the fire tube. And the fire tubes.. got them to last a year, at least. One of the things called for ... was that we pull the treater apart once a year and inspect it. Sometimes, you'd find everything perfect, and three weeks later you'd have a hole in that same treater ... we had fairly high H_2S and we had a little CO_2 ... both acid...

An assignment which nearly cost Hugh his life involved the removal of wax in the flow lines and the use of pigs:

> ...It was at Redwater Battery 4, east of town ... we'd been having problems with some pig traps. In fact, one pig that we had air-discharged - used air to blow it through a new line with - didn't even stop for the pig trap. It took the pig trap and all and went through a doghouse. And an operator was sitting there - on one of these dunce stools - that we used to call (them) where we make reports on the pig - and it went right between his legs and out the other side of the wall...
>
> One that I was working on ... had fluid in it ... We were pumping it at the other end with a go-devil pump - an ordinary gas motor-driven pump and a tank truck - and the pig came through.. It was a brand new pig trap.. and it didn't stop for the end of that pig trap. It hit what we call a 'line blind' in there, and the line blind broke. We think that is what hit me right between the eyes, here.. and they hauled me in to Redwater.

Treater fire tube with sacrificial zinc anodes. L to R: Bill Cousins of Imperial Oil and Jack Godwin of National Tank.

Furnished by W. J. Gibson

Ernie Starr, who used to own Starr's Ambulance here in Calgary, happened to be giving First Aid lessons there ... patched me up and got me in to Edmonton. They said, 'You'll be just a couple of days.' They operated on me, and I got 20 silver clips in my skull and 36 stitches! One of the knives on the go-devil took one layer of my skull.. just like a razor slice through there ... The biggest problem was that there was so much wax and sand under my scalp, they had to clean it out several times before they could patch me up ... 95 days later, I got out. I can tell when the weather's going to change ... that's my barometer ... They kept me in the office for a while, because of the cold weather effect on me...

The service rig industry had its share of difficulties, as Mackenzie explains:

...We started having trouble with all these wells dying. Kenyon's had one rig on, swabbing - every day of the week, we'd swab at least 20 wells back into production. And Prairie Well Service, which was Oly Edmonson - he had one on there, a Cardwell hoist, and a guy by the name of Cox had a rig in there. That was the first three service rigs in there besides Keith Dawson's spudder.. from Lloydminster, and we'd do completion work with it ... In December of '50, with 17 service rigs

- all the service rigs in Canada were working in Redwater - we converted in excess of 100 wells from flowing to pumping in one month...

I can remember, we had a big blizzard come there, and the Assistant Foreman for each area was responsible to keep the roads to the wells open. I would give them a list of which wells would be next to go on pump, and they were responsible to have the leases all clean. They turned this one snow-plow operator loose one night - gave him a list of all the wells he had to plow - and didn't tell him all these sucker rods were laying on the ground ... he ruined about 20 strings of sucker rods in one night - drove over them with a cat! There was a little bit of fur flying...

Bucyrus-Erie model 36L service rig (not self-propelled) owned by Keith Dawson, the first in Redwater.
Furnished by H. MacKenzie

Hugh formed 4M Service Rigs (Management, Machines, Mobility & Manpower). He enjoyed the challenge and was always looking for new methods of doing things. He developed the first truck-mounted and spooler units, that were unique to Canada, and the first automatic banding systems for REDA cables -all related to REDA pumps - there were so many of them going in at the time. Hugh finally got out of service rigs in late 1978, with no regrets.

Arie van der Lee, as the name indicates, is of Dutch stock. His father immigrated to Canada before WWI, settling near Chipman, east of Edmonton. It was there that Arie first saw the light of day in 1924. After a childhood spent on the farm, he proceeded through Grade XII and, in 1943, decided to take math and physics at the U of A, signing up for First Year Engineering.

Hugh MacKenzie standing before the author's Redwater map.
Furnished by Author

Arie reported for his first job to Maurice Paulson at Devon, who said, "Go and test this well". "I'd never been close to a well ... I didn't even know what a pitot tube[10] was." Later, in 1949, other opportunities beckoned with National Tank, a supply company located in Edmonton (104th Street and 66th Avenue). He was hired by Jack Godwin, to whom barbecue outfits were just as important as production equipment:

> ...We built all these fire pits and warming closets ... Jack drew up the designs, and we had the welders in the back of our yard in Edmonton hammering that stuff together ... I'll tell you one thing, they were hell for stout. It took a pretty good sized truck to move 'em around ... He did it (barbecuing) in quite a different way, with pitchforks, where you were turning the steaks and ribs, slopping the barbecue sauce on with wiping rags tied to forks ... kind of 'in the rough'...

According to Arie:

> ...We were big in all types of production equipment. Most of it was bolted tanks, welded tanks, gas/oil separators and emulsion treaters for crude oil dehydrators. And we kept a very large stock of treaters,

10 Orifice which is placed in the stream of gas flow. The gas pressure forces mercury down a calibrated a U-tube, and the displacement reading is converted to MMcf/d.

Arie van der Lee, vice president of AFS.
Furnished by A. van der Lee

> tanks and separators so that, if somebody felt they were getting a well, or the indication they had one, by the time that well was completed, it was our job to have that production battery ready so that they could immediately start producing. One day lost and everything hit the fan...

I remember Leigh Constable calling me in - I was running the branch at Edmonton - and he said, 'Can you deliver 27 treaters before break-up?' And, of course - trying to be a salesman - I answered, 'I should be able to do this.' But it proved to be quite a problem, because the break-up came quite early ... we were running out of Edmonton ... on every back road we could find, for a while, to get these treaters delivered...

Initially, everything was fabricated in the U.S. Arie describes how staves, shipped from south of the border, were assembled into tanks.

> ...Each stave was 5 ft by 8 ft, weighed approximately 190-200 lbs depending whether it was galvanized or plain, or 12 gauge or 10 gauge steel. A crew of four men (farm boys) could build a 1000-barrel bolted steel tank in one day. There were 28 200-pound staves in it and 10,700 half-inch bolts. And all these bolts had to be washered. And those guys! Some of them would carry two of these staves on their backs! Nowadays, to build a 1000-bbl tank, so it wouldn't leak, would take a

crew of 6-10 men a week compared with how those guys used to work ... They never even thought about ... a break...

Bolted tanks were far better than the welded.. because they could come up galvanized. And Redwater's oil and gas and water are pretty corrosive, so we'd build galvanized steel tanks. And then we introduced a new concept when we put aluminum decks on 'em, because the hydrogen sulphide and the moisture wouldn't attack the aluminum; whereas, it would eventually eat up the steel ... and a lot of them are still there ... Then we went a step further - we built 'em with coned bottoms, with the sumps, so that the corrosive water could be easily drained off ... to keep it from puddling and lying on the bottom of the tank...

As for treaters:

...They were ten tons ... fifteen tons. And they'd come up in one piece - from the States ... 4-ft, 6-ft, 8-ft, and 10-ft - so you'd have them.. all approximately 30 ft high ... and they would stand vertically... Instruments and controls weren't as reliable as nowadays ... that's the reason for having them 30 ft high, so it could operate at atmospheric pressure if you lost the pressure off the vessel; whereas, with a horizontal treater, you need to hold the pressure on it or you'll flood and go to the flare pit...

One of the first treaters we put in at Redwater was for Imperial Oil, north of town ... an 8-foot treater ... and it operated for six weeks when we got a panic call: 'Treater is leaking!' We went out there and ... it was! The fire tube had corroded, the furnace eaten right out. So, we got hold of all our good Imperial engineers, and sat down and figured out what should we do. We felt the answer would be to put in sacrificial anodes. We welded bolts onto the fire tube, put these zinc washers on it, stuck it in, and that prevented the tube from corroding out! Those were the first anodes that had been used in treaters ... some of those vertical treaters are still standing ... So the answer was zinc anodes, without any measurement or any resistors, and periodically check the fire tubes. The one big service job now is pulling the fire tubes and checking them, and replacing the zinc anodes...

The standard treater has a large free water knockout compartment in the bottom, it could handle a considerable amount of water. But we'd get into the really large volumes, like the big field gate treating station that's at Redwater now. We put in large flow splitters, free water knockouts - very large vessels with a big horizontal surface area for allowing separation of the water and the oil - before we'd go in to the treaters. The large treaters, that are used for the field gate batteries, are horizontal using the chem-electric grid system. The alternating

> current helps excite the little particles of water ... coalesce the water, and form big enough droplets so it can separate more readily...
>
> ...Before people would install a treater, sometimes they'd produce their oil, and they'd have the oil rejected because it was cutting too much water . .. then we'd take out a portable heater and batch-treat this oil. We were batch-treating - it was minus 50. I think that one winter was about the coldest going. We'd switch lines, the lines would congeal. Then you'd take the discharge line and switch it to the section. Oh, it was terrible! Thinking back now, I just wonder how one could go through something like that...

National Tank claims to have contributed to a new method of production at Redwater:

> ...We knew that the wells would be producing water ... to be removed from the oil, so it required emulsion treaters. We would install them, and tanks and separators, and have a circulating system to keep hot oil pumped in the tank, if necessary ... avoided the necessity for the extra cost of steam boilers ... Some of the people at Imperial took a lot of convincing before they would get out of the idea of using a boiler, but they could see the advantages ... Treaters were a fairly new concept. But one thing that really made National Tank a successful company was their design of emulsion treaters...

Gradually, National Tank did more and more work in their own yard and shop. In 1960 they went into the business of manufacturing, and Arie moved down to Calgary to help build the shop on 58th Avenue and 6th Street S.E. By 1974 there had been several changes in the organization, the outcome of which was Arie's setting up a branch of Crest Engineering, a division of Combustion Engineering:

> ...I set up my own company, which was Global Trading for the purpose of exporting, because the industry had pretty well matured then. It had changed. It was reasonably successful for quite a few years...

Van sold the company to Lavalin. Global Trading then became involved with Creusot-Loire, with Lavalin projects being conducted in the permafrost in the northern Komi Republic (Soviet Union). His latest venture is the conversion of diesel engines to run on natural gas, with contracts in Korea, Japan, Hungary, Iran and Malaysia.

Rudy Roloff, now retired in Redwater, still does some farming south of the river. Unlike many of his countrymen, Rudy was a relatively late arrival from his native Poland. He was born in 1927, and his family landed in Montreal in 1928. Waiting for them was his grandmother, who had already settled about three miles west of

Bruderheim. Rudy grew up on a farm at Redwater, attained Grade XI, then went to winter bush logging at $1 an hour and peeling logs at 2¢ a lineal foot. One day, after returning to the farming, he was driving by a Seaboard well and got to talking with Bob Bischoff, their field superintendent. Bob suggested that Rudy "start tomorrow morning" as a "Roustabout." This was September 20, 1950, at 97¢/hour; a few weeks later it was $1.07/hour.

Rudy recalls the set-up there: BS&B tanks with a Maloney-Crawford treater. Seaboard had been the first to use the submersible REDA pump: 50 H.P motor, and the tubing - with the cable - was lowered to the bottom of the hole. The high line of 13,200 volts was fed into a transformer which cut it to 440 volts. It was this power that ran the motor. This 40-ft unit impeller pumped 2200 bpd of fluid, cutting 60% water. When a well watered out, they ran a Model D permanent packer, mixed up 200 pounds of special cement (goop), pumped below the packer, pulled out the stinger and landed the bottom of tubing above the packer to reduce the water cut. Much later (1958), when Seaboard was taken over by Texaco, the latter's policy was retrievable packers - so they could use them again - typically Texaco.

Prior to the takeover, Rudy had been elevated to foreman and occupied a house on the riverbank, a freehold tract. When Texaco took over, his Seaboard pension plan was put in the deep freeze and wouldn't pay off until he reached his seniority age. Texaco's pension plan started off right away, and he was given the same seniority (so he would have his holidays), but Texaco contributed only 3% vs his 6%.

In 1959 Texaco gave Rudy more responsibilities, sending him out all over northern Alberta on drilling projects. He reported to Vern Lindaberg, Texaco's office at 10305 Princess Elizabeth Drive. Getting around in the field was difficult because they had to use the ferry at Vinca until 1967. During river break-up, a cage holding four men was pulled across by a motor. When asked about contacting the different batteries (no phones), "You just went to see them."

He remembers the gathering of Redwater solution gas wherein Majestic Contractors brought lines in to the lease, and the only measurement of the gas was on the charts. He also describes the block allowable, where a producer has to prove a certain well would make its allowable, but you transferred that to other wells inside the tract - a full allowable, if continuous. When Rudy wanted to retire at age 55, Texaco imposed a penalized annuity, which he accepted. He left on January 1, 1983. During all the time he worked in the oil patch, he and his brother-in-law had farmed 290 acres, covering the work to be done on week-ends and vacations. "I'm glad I had the opportunity to work in the oil fields, but when the time came I was ready to pack it in." Myrt, whom he married in 1952, kept the home fires burning. Today, Myrt drives the truck and Rudy runs the combine. They never really left the farm.

The **Cormack** name is well known in the oil patch. George Sr. was on the Leduc discovery well and had a long career in drilling. Brother William was a career bush pilot, and brother Si (official name: Alexander) was into production and later operated service rigs.

Si was born in India, the son of a British Regimental Sergeant Major who emigrated to Cabri, Saskatchewan in 1904. Si worked his way over to High River, Alberta, where he met his wife, Mildred. He spent some time in Norman Wells, on the CANOL Project, and recollects many adventures there, as does Mildred. She was an R.N. by then and vividly recalls helping deliver the first white baby[11] born at Norman Wells.

Si was transferred to Redwater in April 1949 to start as Production Foreman under Bill Cousins and worked there until he and Imperial parted company. Walt Dingle called Si into the office to give him his package. "Best thing that ever happened," said Mildred. At this juncture, Si decided to get into service rigs.

> ...They wanted me to take a service station over in South Saskatchewan. I didn't think a service station was very good, so I got the rig instead. We had $2000 in the bank when we - you could say -parted company. And I told the bank to hold my Imperial shares - I didn't want to sell 'em - and borrowed this money on it...
>
> I bought it from Oklahoma. Coopers had it in Saskatchewan - they sent it down to Oklahoma and fixed it up the way we wanted. When it was ready, sent a man down to drive it up. And that was it, we started the first of May, 1960.

Gordon Cormack, Si's son, works out of a large Atco trailer - on Shkolny's Sec. 19 - a last vestige of Gulf's Redwater operations. He not only operates the original Gulf leases but also the Royalite Triad and old B-A Crown Reserve parcels. Gord looks after 107 wells, 29 abandoned, and one under the Town park which it is no longer economic to rework. Costs have shot up to where Gulf's electric power

---

11 The baby, Howard Dingle, is now one of the senior executives at Imperial Oil.

Si and Mildred Cormack

Furnished by Author

bill is $2 million a year, so the operation is approaching its economic limits. In 1980, the maximum was 80,000 bbls a day of crude. Even then, water had reached 600,000 bbls. a day. Today, production is 8200 barrels of crude and one million barrels a day of water.

With three runners out in the field, the trailer is vacant most of the time. On the wall are group photographs of happier days with Dan Motyka, Harry Carlyle, Ed Lakusta and other notables. Gord is the last of a generation of oil men, attempting to salvage whatever he can from the wells that are now producing 99% water.

In a nearby shack Diane Dowhaniuk works half days on a contract which includes Gulf, Talisman (old Home Oil acreage), Petrocan and Imperial Oil. She analyzes samples and calculates the "oil cut". Because Imperial is the largest producer, they have established what they call a three-phase system which is completely automated (separate readings for gas, oil and water). The other operators are on

Diane and Gord, with a flask illustrating a dying field - a tiny amount of oil in the neck.
Furnished by Author

a two-phase system wherein gas and oil are carried in one line and water in another. Diane and Gord are two of the many custodians ministering to the dying giant.

Gordon Teske started as a truckdriver, hauling crude to the hastily erected loading rack at CNR's Kerensky[12] siding for both Imperial and British-American - the tank cars were marked to prevent mix-ups. There were two men to each vehicle, hauling night and day. The basic unit was the three-ton single axle holding about 55-60 barrels. Gordon remembers a period when B-A's Calgary refinery was so desperate for crude - the market price was dropping - that they trucked it directly from the wellhead. They made at least 52 round trips to Calgary, each trip taking up to an entire day. They had to go through the Edmonton traffic, most of the small towns on the way down, not to mention the traffic through Calgary.

Gordon eventually felt he wasn't being paid enough and went to work for Imperial Pipe Line at 86¢/hour. Floyd Madill[13] later lured

---

12 Some foreman with Communist leanings must have chosen his hero's name for this out-of-the-way siding, never dreaming that capitalist oil firms would exploit it. Alexzandr Fyodorovich Kerenski (1881-1970), Russian revolutionary leader, was premier of the USSR July to Nov. 1917. He moved to the USA in 1946.

13 Madill started with an Imperial Oil gravity crew and later headed up Peace Pipe Line.

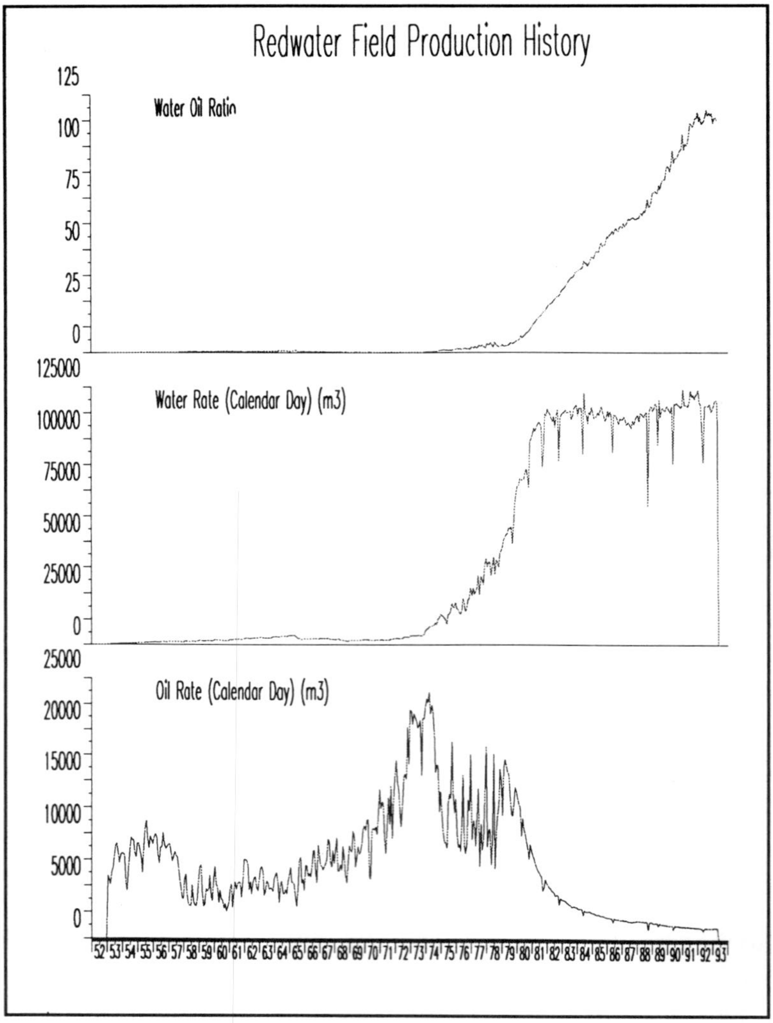

Redwater Field Volume History, 1975–1992.

Furnished by D. Carsted

Gordon, and others, away to set up Pembina Pipe Line (a Mannix operation). He worked there for nine years before returning to farming in the Ryley area.

Kerensky's loading racks, handling four cars each, soon proved to be inadequate. Then there was the congestion in the Redwater town siding - so many trucks were scattered around the tracks that the trains couldn't get through. Furthermore, the Redwater grain elevator people objected. They said it was not safe to ship crude out of "their" siding. In 1949, Imperial Pipeline approached the Pasemcos to purchase a 50-acre fraction of their land (SE/23-57-22 W4M) along the CN right-of-way. They intended to install super-large tanks (100,000 bbl each) and a loading rack capable of handling 30 cars at a time. The Pasemcos sold the parcel for $200 an acre, cash. Victor Pasemco recalls that the facility was to be called Redmont, but the name was never adopted.

Jack Harvie, horse fancier, poker player, golfer, and one of the best story tellers in the oil patch, has a keen engineering mind focussed to finance. No wonder he advanced to a top position with Imperial.

Jack, a Civil Engineering graduate of the University of Manitoba, was involved in all aspects of engineering with Imperial, including oil transportation. He understood the difficulties of the latter and the resulting need to turn to pipelining:

Jack Harvie, always ready with a story, February 1993
Furnished by Author

> ...I took Whitney Sexton - he was manager of Imperial Pipe - up to get them planning how to get this oil out. They ended up building tanks to the west side of town. We fed those tanks, and they took it from there. Once the directors in Imperial realized what they had in Redwater - they knew the local market couldn't handle this oil - they needed a pipeline. So they sold their interest in International Petroleum to "Jersey" to raise the money to get their 30% share in Interprovincial Pipe Line...

Interprovincial now boasts the longest pipeline system in North America: 3700 miles. Who would have thought that a giant system would emerge from a puny oil line of the thirties that moved crude from Turner Valley to Calgary?

The Leduc discovery provided too much oil for the Alberta refineries but not enough to pipeline outside of the province. The Redwater discovery changed all that, eventually yielding three times as much oil as Leduc. The question became: Where was the market? Ontario now appeared to be the logical target with a refining capacity of about 100,000 bpd, being supplied by the U.S. mid-continent.

There was no statute on the books to regulate pipelines except for the obsolete 1907 Electricity and Fluid Exportation Act, which was all-encompassing in its definition of fluid: petroleum, natural gas, water, or other fluid whether liquid or gaseous. Section 3 called for the licensing of exportable substances while Section 5 required quantity of export to be limited to surplus after meeting Canadian needs. Section 10 provided for export duties of 10¢/1000 cubic feet. Its legislation had no relevance for a brand-new situation: interprovincial movement of hydrocarbons.

The federal Pipelines Act was proclaimed in 1949, and its administration was placed under the Board of Transport Commissioners (all railroad people) as the most appropriate forum before which to deal with this form of inter-provincial trade. At the same time (April 30, 1949), Interprovincial was incorporated as a Special Act Company. Two of its directors, O. B. Hopkins and J. R. White, were Imperial Oil senior officers. Sweeping powers were conferred on the company - ranging from building town sites, leasing blocks, and selling land to the loaning of monies to employees to occupy these houses.

The immediate goal was to pipeline crude to Superior, Wisconsin and then tanker it through the Great Lakes to Sarnia during the navigation season rather than trying to blast across the Canadian Shield. Hearings were conducted in 1949 and approvals given to construct from Edmonton to Gretna, Manitoba, on the international border. The portion from Gretna to Superior required no U.S. approval. Financing was in place in September 1949, $90 million in bonds having been promised. The line was officially opened October 4, 1950, by Messrs. G. L. Stewart (president, Imperial Oil), Hon. E. C. Manning (premier of Alberta) and Hon. C. D. Howe (minister of Munitions and Supply).

The first part of the system was 1180 miles long. Twenty-inch pipe was procured, and contracts for construction were let in November 1949. The original terminus was Edmonton. but during the first year it was decided to extend the system from that city to Redwater, a distance of 29 miles. The first oil was turned into the line at Redwater on August 25th, 1950, from the Pasemco terminus. This enabled Imperial Oil's production to peak at 86,000 barrels/day in 1952.

It was soon necessary to extend the line to Sarnia to handle increases in demand, once more denying the harsh reality of the Canadian Shield hard-rock by cutting through northern Michigan and crossing the Straits of Mackinaw. Yet another connection to the Sarnia area was achieved by building a loop via Chicago.

Interprovincial's and Imperial's 50-acre tank farm dominated by two 100,000-barrel tanks. Redwater is in the distance to the northeast.

Furnished by V. Pasemco.

Redwater was only the beginning. Interprovincial diversified and extended its operations into condensates, synthetics (from McMurray) and heavy crudes - all batched. Later, through a participating company, they built a 12-inch 540-mile line connecting Norman Wells to Zama in northern Alberta.

The Redwater tank farm is now long gone, as is the ownership of the pipeline, now incorporated into Alberta Energy Company's crude line from Fort McMurray. The 50-acre Pasemco fraction was turned over to the Town of Redwater for $1 to provide a recreational park. Victor Pasemco offered to buy it back, but his letter to Interprovincial went unanswered!

Not to be forgotten in the whole engineering infrastructure was Imperial's addition to Redwater to provide housing and offices. Gibby Gibson describes:

> ...They purchased a bunch of land on the north side of Redwater and laid out a town site ... a residential area. There was no thought of putting up any businesses or things like that. And within a couple of weeks they'd moved in an office-bunkhouse accommodation which was the old army H-hut type. Two wings were bunkhouse, and the other two were for the offices...
>
> I was in one of the wings. I guess there would be 20 of us in one, beds side by side and a little stove in the middle. A little gentleman used to come in and fill it up with oil once or twice during night, depending on how cold it was. And they had toilet facilities installed. But the water! You couldn't drink it! So we drank bottled water brought out from Edmonton. That's where we spent the winter of '49...
>
> While all this was going on, they were busy constructing some proper bunkhouses, two-to-a-room type with a lounge in the middle ... a very adequate field office for us, and warehouse, pipe racks ... We also put up a smaller building to accommodate three or four girls that were working in the office for us...
>
> They showed a lot of foresight when they laid out the town. They only built on every second lot. There was a marked sameness in many of the houses. Their thinking at that time - and this has certainly been the case as other people move into the community - they will build in between and have different style homes, so you won't get the sameness. And the cost of building them will simply be a tie to the utilities ... By the early summer, we had all the amenities of home: gas, light, water, sewer...

Ken Oakley, townsite engineer for Imperial, provides more detail in a report that is undated but appears to have been written in 1952. He describes the site, adjacent to the hamlet of Redwater, as comprising approximately 100 dwellings, offices, warehouse, garage and staff quarters[14]. When subdividing the townsite, consideration for existing wellsite locations had to be built in. The length of any block could not exceed 600 feet. Oakley comments on the decision to leave alternate lots vacant. It resulted in an untidy appearance from the growth of weeds, but that is history now - forty years later - mature manicured lawns and treed streets greet the visitor. The homeowner was required to complete the levelling of the top soil. Sidewalks were included in the specifications. However, according to Jack Harvie:

> ...Curran made the decision that it was just an oilfield town, and there was no money to be spent on sidewalks. Well, you know, it was pretty muddy up there, and we decided this was impractical. So, we went ahead and put in sidewalks - budget or no budget. And when he came up, he said, 'I thought I ordered no sidewalks!' I said, 'Well, these are all young women with babies. They need some place to wheel their baby carriages.' That's all I said. He never bothered me again...

The house specifications called for extra charges for thermostat control. Imitation shutters were to be painted to match the color of the roof. All exterior woodwork was two-coat, fume-proof, non-leaded paint - even then, there was concern over lead poisoning. The bottom panel of the bathroom door was to be replaced by a grill; presumably there was no window. Fir flooring was extra over the basic masonite which was "laid according to manufacturer's specifications, left clean and ready for waxing." Oakley reports problems with acceptance because workmanship did not conform with good construction practice. The shortage of skilled help (and accommodation for them) resulted in substandard work.

"Dwellings were constructed on a lump sum contract basis. A single contractor (Bird) built approximately 90% and there were three designs." Design limits of dwellings were governed by the stipulation that the maximum monthly mortgage, utilities and tax payments could not exceed 25% of the employee's salary. A down payment of approximately 10% was required. Gordon Teske, Imperial Pipe engineer, purchased a 500 sq. ft. "Bird house" for a total cost of $4800 with a $35/month mortgage payment.

---

14 Oakley does not mention the Company's efforts to consider Egremont to the north as an alternative townsite.

The basic design of a one-entrance, single storey 20'×22' dwelling was submitted but this was improved by increasing the size to 22' × 24', adding a porch, setting the building on concrete posts and framing a dugout underneath. Another design incorporated the high basement ceiling permitting the building of extra rooms in the basement. Fifty of this type were built. Two other designs were used: seven one-and-a-half-storey houses and two bungalows. (The latter two, one of which was Bannantine's, were constructed by E. Morin.)

Oakley touches on the "haves and have nots" when he says:

> ...The advantage of not having to set up a business district has been outweighed by the poor public relations which resulted from the community having half of its area supplied with water and sewer facilities while the older half lacked such facilities...

Despite extensive subsurface investigation, no shallow water was available, and that meant building a line to the North Saskatchewan River. The small island located at the confluence of the Redwater River was chosen as the intake. Sewage disposal was about one mile northeast downstream. Oakley provides a technical description:

> ...The water brought up from the North Saskatchewan was a composite of high mineral and low oxygen content with low pH. Corrosion was so severe that a heavy deposit of iron rust settled out on domestic utensils. The problem was solved by the injection of sodium hydroxide. The pH was upped to 8.5 and the fact that there was this infiltration gallery on the island ensured the purification of the water (proved by B. coli tests) even though it contained raw Edmonton sewage...

Lauder Nowers, Imperial personnel officer, describes the water in more "folksy" terms:

> ...One fellow said it was the stuff that Christ walked on ... You can't believe, it was so hard. Being a bunch of teetotalers, we noticed when you poured some booze in a glass and added water in it, it immediately turned black. It was the most revolting drink you ever saw ... The girls who were washing diapers ended up with a whole line that dirty khaki colour." Then Jimmy Young[15] came up with his bag of tricks ... treated it with some alum additive ... finally got it so it didn't turn the booze black, and the girls could do a washing that came out the way they expected...

Reservoir determination, water disposal, production, transportation and town planning were all an integral part of Redwater's

15 Imperial Oil's non-degreed yet highly respected technical guru in charge of research.

Some dwellings were more equal than others. Bannantine's house was (is) in a preferred location at the northwest extremity of the townsite. Mildred Cormack recalls "...Lots of windows ...I thought that was the biggest house..."
Furnished by Author

Bleakley's Bird House (500 sq ft) with small stoop (extra), no back door.
Furnished by S. Bleakley

engineering infrastructure. An aspect that Imperial was not particularly interested in was gas gathering; they claimed it was uneconomic to gather and process. During the first few years, solution gas was flared at the batteries. At the same time, Imperial was aware that there was a need to process out sulphur and to reduce air pollution - they retained Brown & Root to design a plant for this purpose.

A cornerstone of Conservation Board policy was the prevention of waste. This was a contentious point with regard to Redwater gas, especially since the reservoir was under-saturated: there was less than normal amounts of gas which could be dissolved in the crude at reservoir conditions - unlike Leduc, there was no gas cap. However, the Board felt there was an irretrievable loss of natural gas, as well as liquids (pentanes, propane, butane, isobutane), all of which could generate royalty for the Crown, if marketed.

By 1955, the Conservation Board had warned Imperial that their crude production would be shut in unless they agreed to gather the gas. Despite Imperial's protestations (or "good negotiating strategy" as Ed Lakusta put it), a hearing was held in September of 1955. The following item from the Daily Oil Bulletin of November 1, 1955 tells of the outcome:

> ...Construction of a $3,000,000 gas processing plant to handle gas obtained with oil production from Redwater field will be completed by the end of 1956, it was announced today by W. B. Dingle, Edmonton division manager for Imperial Oil Limited. Brown & Root Ltd., Calgary construction contractor, will build the plant, and Majestic Contractors Ltd., Edmonton, have been awarded the contract for construction of the pipeline gathering system. Both these contracts were awarded on the basis of low bids...
>
> The plant will be erected on a site one mile east of the town of Redwater. It will be capable of processing 10 million cubic feet per day of Redwater field gas and main products it will manufacture will be butane, propane, and pentanes. Major construction is not expected to start until next spring although one pipeline crossing of the Saskatchewan river may be completed during the winter freeze-up. The plant will go into operation by the end of 1956...

An Imperial inter-office memo stated that the gas would be gathered from almost 800 producing wells and the volume, about 85%, would be delivered to the plant which had been designed for

10 MMcf/d with an input pressure of only 10 psig. The official date of plant start-up[16] was October 26, 1956.

At first, the residue gas was piped in a six-inch line to Sherritt Gordon's nickel plant at Fort Saskatchewan. "It removed the glow in the dark ... resulting in a peaceful environment" (Ed Lakusta). Later, gas went to N.W. Utilities. Sulphur was shipped to the Fertilizer Plant.

In 1970,[17] when a fieldgate battery was installed to replace the many outlying satellites, a two-phase system was introduced whereby the gas and the crude would be brought in together to be treated in this large facility.

Gas Plant located in SE/29-57-21.

Furnished by Author

Plant capacity was doubled in 1971 to 22 MMcf/d because of field production increase and diversion of Nipisi crude into the facilities. Gas from dead oil batteries was compressed and delivered to the main plant under a separate arrangement. Water had increased to the point where free water knock-outs were installed in the field and the water piped direct to salt water disposal wells. Today, the

16 See Appendix XIV for list of start-up personnel, supplied by John Lang.
17 Appendix XV gives an Imperial Oil summary of Redwater in 1970.

Jim Hawkins, Area Manager for all small oil, gas and NGL operations in western Canada. The photo on the wall behind him depicts Esso Rig No.7 on a caisson-retained artificial island in the winter of 1984-85 in the Beaufort Sea. Jim engineered the design of these islands.
Furnished by Author

John Lang in his Redwater home.
Furnished by the Author

plant operates at a much restricted capacity. When will it be uneconomic to continue?

John Lang, who was with Imperial at the start-up of the gas plant and later with the butane cavern near the Fertilizer Plant, lives in Redwater. His remark, concerning the field's low profile considering its contribution to Imperial's coffers, is cogent: "Redwater was always forgotten, and yet Redwater paid for a lot of fields."

The engineering emphasis today is focussed on environmental matters: abandonment and cleanup. Redwater started off with too many wells. The authorities did not realize that 160-acre spacing would have been more than ample to achieve maximum recovery. Think of all the production casing and pumping equipment that could have been eliminated. But much more importantly, consider the unwanted and unnecessary wells which are now inactive. They are either suspended or in some form of semantic cover-up while awaiting a solution to the problem of abandonment.

Frank Manyluk, who officially resigned from the Conservation Board in 1971 and stayed on in an advisory capacity until 1982, acknowledges this transformation:

> ...If I had to go to work today, I'd have to reconsider ... The majority of Board work is involved in environmental matters rather than drilling production reservoir problems and the like. And, although I realize environmental matters must be dealt with, it's not my cup of tea...

It is neither the landowner's nor the producer's cup of tea; each has a price to pay. The Conservation Board is faced with producers who are putting off the evil day of spending money to abandon wells and farmers who see a "diminishing crop" of a different kind - the "cash crop" from rental of lands for wellsites and batteries (regardless of their status).

The Conservation Board had the foresight in the mid 1980s to establish an abandonment fund of $3 million to be shared 50-50 with industry. The interest from this fund has been used to abandon some wells. In 1989, the Board took further initiative and presented recommendations to industry, resulting in the creation of the Orphan Well Steering Committee in 1990. An administrative subcommittee was set up to establish and maintain an industry fund. Assuming 29,000 inactive wells, a levy of $60 for each of these will generate $1,700,000 for fiscal year 1993-94.

Bill 23, officially known as "The Environmental Protection and Enhancement Act" places the onus on company officers and directors. The precedent for this legislation is the landmark case of Northern Badger, wherein that company defaulted and went into bankruptcy. The ERCB directed that the wells be sold to a new owner and the selling price for all (with the exception of seven wells) was $226,000. Then the ERCB ordered these seven wells to be abandoned at a cost of about $200,000. One Northern Badger creditor, Panamericana de Bienes y Servicios S.A., was forced by the Board to pay for this work. The Supreme Court of Canada turned down Panamericana's appeal.

The Canadian Institute of Chartered Accountants now requires companies to include land clean-up costs in their corporate accounting. And bankers are voicing concern. In one case (not Redwater) five banks walked away after collecting only 15 per cent of the monies owing them to avoid the responsibility for cleaning up a property. Overall the fallout from this case could entail as much as $1 billion. Access to capital could be greatly restricted if banks are also held responsible for the environmental problems of their clients and, as a follow-up, financing within the oil industry itself could be much more difficult. Ian Doig, in his article (Doig's Digest, August 1992) refers to this state of affairs as "a ticking time bomb."

6 THE LAND: FREEHOLD

No Freehold Minerals for the Homesteaders

The term "freehold," for purposes of this narrative, means absolute ownership of all mines and minerals (except gold and silver) by an individual or corporation. Land tenure in the Redwater Area evolved in such a way that no farmers obtained their mineral rights with their surface acquisitions (either filing or purchasing). How did this happen?

The Dominion of Canada (Maritime Provinces, Ontario and Quebec) came into being in 1867 with the stroke of a pen.[1] The balance of this vast Western Canadian wilderness, known as Rupert's Land, had been deeded to the Company of Adventurers, the Hudson's Bay Company, by Charles II in 1670. In 1869 the new Dominion purchased this territory (surface and minerals) for the sum of £300,000[2] having been bankrolled by the Mother Country, Great Britain. The bulk of the land became the Northwest Territories. The deal excluded existing "factories" (trading posts) and 1/20th of each township (Twp) south of the North Saskatchewan River within what was defined as "the Fertile Belt". Thus, what would become Hudson's Bay Oil and Gas Company Limited surface and mineral acreage was acquired on the basis of Sec. 8 and ¾ of Sec. 26, approximately 1.8 sections. Where the township number was divisible by five, all of 8 and all of 26 were included to make up the difference.

The Federal Settlement Act of 1872 enabled newcomers to file on this Crown homestead land. Requirements included the building of shelter, clearing and cultivation, and one's intention to become a British subject. (No such thing as a Canadian then!) Title to the surface and minerals would then be conveyed to the homesteader.

Some functionary in Ottawa thought the sod-busters were getting too much, so on October 31, 1887, Order-in-Council PC.1070 reserved on crown lands west of the third Meridian[3]

> ...to Her Majesty, Her Successors and Assigns forever, all mines and minerals which may be found to exist within, upon, or under such lands, together with full power to work the same, and for this purpose to enter upon, and use and occupy the said lands or so much thereof and to such an extent as may be necessary for the effectual working of the said minerals, or the mines, pits, seams and veins containing the same...

1 Manitoba (SE portion only) entered Confederation in 1870, PEI two years later.
2 See Appendix XVI for comments on Hudson's Bay Company's Deed of Surrender.
3 Runs north-south through Central Saskatchewan.

This meant that homesteaders would obtain title only to the surface. Jack Dallas, Turner Valley promoter, was reported to have said "Keep the minerals away from the settlers in the original deal." Consider the wealth that would have accrued to the homesteaders if this move had not taken place. It has been argued that this withheld distribution of minerals would ultimately benefit the populace as a whole instead of just those lucky enough to have acquired minerals with their titles. Sales of Crown Reserve by the Alberta Government is the best example of such a benefit.

Administration of mines and minerals, therefore, became a federal responsibility resulting in the evolution of statutes directed mainly at hard rock and coal. Oil and gas regulations, primitive at first and never really effectual, were handled from Ottawa even after Alberta and Saskatchewan became provinces in 1905 - to everyone's detriment. This continued until October 1, 1930. When it came to hydrocarbons, one can only describe the process (or lack thereof) as benign neglect. The irreversible ruination of the Turner Valley reservoir resulted from ignorance of basic petroleum engineering principles. The lax Ottawa attitude was best exemplified by the then-Minister finally declaring on January 1, 1930, after years of failure to collect royalties, that they would now be payable on Turner Valley production.

The populist movement of the 1920's fought hard to wrest minerals from the Feds to no avail. In 1926, the Alberta Government passed a statute which was intended to control all operations, both crown and freehold, but no regulations were established thereunder. It was not until March 28, 1931 that the new Alberta Oil and Gas Wells Act was proclaimed, entitling Edmonton to administer all oil and gas operations, crown, non-crown, drilling, production and the collection of royalties.

Alberta's Petroleum and Natural Gas Conservation Board came into being July 1, 1938 under the chairmanship of William F. Knode, hard-nosed petroleum engineer who had worked for the Texas Railroad Commission. The final transfer agreement between Canada and Alberta had been signed on March 5, 1938. Hon. N. E. Tanner couldn't wait for Ottawa approval, so Alberta enacted the statute on April 8, 1938. It did not, however, become law until November 22, 1938. All of this was pursuant to the Natural Resources Transfer Agreement, assented to June 24, 1938 and countersigned by Ernest C. Manning. How slowly do the mills of bureaucracy grind!

Most of the history of how the railways (and their beneficiaries) acquired their freehold titles is to be found in Leslie Munroe's report.[4] "...In view of the magnitude of the undertaking, the Government at one time considered that it would be necessary to set aside an area in excess of 50 million acres..." The first step in this direction had taken place in 1880, when Ottawa gave what would become the CPR $25 million cash and a grant of 25 million acres of crown land (not including Ontario and British Columbia) "in a belt 24 miles wide on both sides of the Right of Way".

In 1889, the Dominion government formalized the give-away of millions of acres, complete with 100% surface and mineral title, to the railway companies in exchange for laying track. Here are selected passages from the enabling Order-in-Council (PC.2167/1889):

Section 2:

> ...The even-numbered sections at the disposal of the Crown are to be held exclusively for homestead and preemption entry unless otherwise ordered...

Section 3:

> ...The odd-numbered sections at the disposal of the Crown and not reserved for or granted to *any* Railway Company are to be held exclusively for sale unless in special cases otherwise ordered by the Minister of the Interior[5]...

Section 5:

> ...The reservation of lands for intending settlers, on receiving satisfactory assurance that the application has been made in good faith and with the purpose of actual residence upon the land applied for...

Sections 11 and 29 in each township were designated School Lands (to provide revenue for education), and those numbered "8" and three-quarters of "26"[6] became Hudson's Bay Company freehold.

This gave the indefeasible right (except for Sections 11 and 29) to "any Railway Company" to acquire fee lands. Railway companies acquired minerals under rights-of-way and station grounds before February 1, 1904, as well as those expressly purchased thereafter.

The CPR was "short" of the 25 million acres under the Land Grant for the building of the transcontinental railway. The amount of this

---

4 Munroe, Manager of Resources, Calgary, wrote a historical report to Mr. Stone, then CPR head of Resources in Montreal, dated April 18, 1959.

5 "No hanky-panky" says Hubert Somerville, confirming that everything was done aboveboard.

6 Lands south of the North Saskatchewan River.

grant was reduced by 6,793,014 acres which was held back in lieu of repayment of $9,880,912 (approx. $1.45/acre). CPR actually acquired only about 18 million acres; they forfeited six million because they had not paid back the $25 million they owed the Government.

They could move out beyond the 28-mile swath, which they did, and there were several alterations; for example, the Irrigation Block east of Calgary and one in the Barons area. Gordon Hawkins said the CPR chose no acreage around Swift Current because that area was deemed un-arable. In order to acquire "make-up" lands, the CPR applied for grants in the Redwater area in 1903 and chose their acreage in 1904. CPR lands were still available to homesteaders directly from the Railway Company as late as 1918, but petroleum and natural gas had long since been reserved.

Dominion land surveys were not completed until 1906 and, according to some records, it was impossible to homestead on any provincially administered lands until after that date. This is at odds with the Homesteaders List[7] which clearly shows settlers filing as early as 1894.

CPR acreage purchased by Leduc farmers prior to 1905 included oil and gas titles although virtually none of them knew of the riches 5000 feet below. After this date, the CPR reserved petroleum, and from 1912 they also set aside natural gas. As mentioned above, records indicate that the railway had applied for their Redwater grant acreage in 1903 and had received title to it in 1904. So why didn't those farmers acquiring surface rights of CPR quarters in Redwater obtain minerals - as did their Leduc counterparts? A thorough check of the homesteaders' filings up to and including 1905 shows that they filed only on even-numbered sections, except for a few quarters that had not been chosen by the CPR.[8] After 1905 it didn't really matter as far as the settler/purchaser was concerned. A few split titles[9] have turned up, but with such a small amount of gas reserves there is little of economic importance in the Redwater area.

CPR soon realized they could not physically handle all their vast endowment, so they "sub-granted" to numerous land companies (mostly of U.K. registration) large tracts of surface and mineral titles which these firms, in turn, would sell off. Remember, their main

---

7 Appendix XVII - Homesteaders on Crown Land, 1894-1905, and their locations.

8 The CPR was the only applicant in the Redwater area; other railways earned fee lands elsewhere on the prairies.

9 Between 1905 and 1912, purchasers acquired natural gas rights by virtue of the petroleum only reservation to CPR. See "LEDUC" pp 138-140 - Borys Case.

motive in disposing of acreage was to populate the prairies and build up railway traffic (and revenue).

The Western Canada Land Company was sub-granted 500,000 acres of both surface and mineral titles in the general Edmonton area on July 7, 1906. This completely unknown enterprise acquired mineral title to all those tracts of land plus certain odd-numbered sections in Twp 56 and 57, Rge 21. These would turn out to be the *real crown jewels*, because they were in the heart of the as-yet undiscovered Redwater reservoir! And how did this "sweetheart deal" occur? We are not sure, but no doubt the conveyancing of the full title took place in a smoke-filled London club room. Later on, somehow and somewhere in these Byzantine dealings, the mineral titles were severed from the surface titles. The latter assets were sold.

But this still does not answer the question of why those farmers who later bought CPR surface rights, on the Western Canada Secs 17, 19, 21, and 31, Twp 57, Rge 21, did not have minerals conveyed with their titles. Just another missing piece in the jigsaw puzzle.

How did Eric Harvie and his Western Minerals Company acquire mineral title to all these 500,000 acres, including those in Redwater? This Calgary lawyer would make a $100,000,000[10] fortune from Redwater crude production.

Eric Harvie was born in Orillia, Ontario on 136 Matchedash Street[11] in 1892 and died in 1975. His father was known as "Dentist Harvie". He married Dorothy Jean Southam in 1919 and they had three children: Donald S., Frederick Neil, and Margaret Joy.

Eric got into the lawyer business in Orillia, articling with Mulcahy in 1911. But, prior to that in 1905, he went with his mother to Calgary to visit his uncles. In 1912 the west cast a spell over him and he completed his requirements for an LLB at the University of Alberta. In the meantime he had joined up with the Canadian Army and served in WWI.

Quoting from John Slatter's "book":[12]

> By accident, H. E. M. Evans, Manager for the company holding the mineral rights happened upon Eric Harvie, an old acquaintance, in 1943. Evans said that he was up against the wall and would have to

---

10 Harvie later displayed public responsibility by creating Glenbow Museum and the Devonian Foundation from this fortune.

11 The author was born on Laclie Street, Nov 29, 1915.

12 The only known copy available is in the offices of Devonian Foundation, Rm 700, 999 - 8 Street S.W. Mr. Slatter has offices in Edmonton, 606 Royal Bank Building, phone 423-4460.

Eric Harvie in the Redwater field, one of his wells in the background.
Courtesy of Devonian Foundation

sell or surrender the mineral rights. Harvie responded, "Don't do that. I'll give you $10,000." Upon his return to Edmonton, Evans[13] reported "I had a good visit to Calgary - made a profit of $10,000 and instructed Mr. Field to proceed forthwith with agreements in conveyancing."[14]

Hod Meech had been an employee of Cord[15] (the automotive genius who built front-wheel-drive cars back in the 30s). He was hired by

13 Evans was the manager for an affiliated company, British Dominion Land Settlement Corporation, but both it and another company named Anglo-Western filed for bankruptcy. In 1943, Harry Evans had concluded that mineral rights would have to be surrendered to avoid mineral taxes which the Department of Municipal Affairs had imposed on freehold titles in 1941 at the rate of 1/3¢ per acre. In 1943 Lands and Mines started to enforce the tax, sweeping each township to locate mineral owners.

14 Gordon Hawkins recalls an acquaintance of his lamenting his ill fortune. Sometime between Eric's purchase (1943) and Leduc (1947) this person was offered a half interest in Eric's 500,000-acre spread for $10,000 and turned it down, not being able to raise that kind of money. If the deal had gone through, Eric would have recouped his $10,000 and still would have a 50% free ride

15 Cord drilled a dry hole near Spring Coulee in 1939 and participated in another failure nearby in 1943.

> Eric to gather up the titles. On May 1, 1944, Hod was appointed Manager of "Managers Limited," one of Eric Harvie's companies.[16]

Eric had paid only $10,000 for the mineral titles, but it was still an outrageous price because interest in oil was virtually nil. Crown lands were going begging, so why should anyone want Eric's freehold?

Conditions in Alberta were improving slightly when, on May 15, 1946, Shell Oil paid Eric Harvie $30,000 for an option to purchase *in fee* (mineral titles, not leases) 300,000 acres: $20 an acre for the first 10,000 acres, $15 for the next 10,000, $10 for the third 10,000 and $5 an acre thereafter. A 2½% royalty was to be granted to Harvie. This option covered what would be the crown jewels of Redwater.

Later that year Shell decided that they could not exercise their option. Several versions of this retreat seem to centre around The Hague's post-war shortage of funds and its Los Angeles office's apathy.[17] Marsh Porter repeated to the author the cable from The Hague to Calgary: "You are precluded from filing on the land." Shell then went to Ottawa to study geology and moved to New Brunswick to explore for hydrocarbons. Only one individual was left behind to "mind the store," Ian Crawford, geologist.

Imperial quickly took up Eric Harvie's option on February 4, 1947 for $250,000 with the secret (not for long) knowledge of Leduc, thus justifying that greatly increased price. But, incredibly, they did not include any of Eric's mineral titles (referred to already as the crown jewels) east of the east boundary of Twp 57, Rge 22. Three reasons for this oversight are listed:

- According to Don Harvie,[18] retired head of Fina, Eric's son and heir to the operations, Imperial did have Western Minerals acreage east of the east boundary of Twp 57, Rge 22 W4M, but dropped it. Gordon Hawkins confirmed that Imperial desperately tried to get back the Western Leaseholds/Minerals acreage that they had presumably dropped.
- Another version, as told by Peter Bediz, had Imperial Oil taking advice from a New York consultant who said, "Don't bother

16 Still in existence - Offices: 7th Flr, 999 - 8 St SW, Calgary.

17 Rumour had it that a memo from Calgary had been gathering dust in one of L.A.'s "In" baskets.

18 Petrofina, a Belgian company, had set up shop in Calgary with Trajan Nitescu, top Rumanian engineer, to be in charge. It purchased Western Leaseholds in 1955, and Don Harvie went over to head up Fina.

Don and wife, Mary, in the early 50s.
Furnished by D. S. Harvie

with any freehold east of the east boundary of Twp 57-22-W4M."

- D. P. McLaws, Q.C., gives the most credible account. "After Leduc was found, Imperial entered into an agreement with Western Minerals to acquire certain of their oil and gas properties. Don Mackenzie was then local manager of Imperial. The area which was to be subject to the agreement was generated in the Leduc area, but nobody paid much attention to the eastern boundary. Mackenzie told me that he drew the eastern boundary, which was a township on the west side of what turned out to be Redwater. *(SAK: This is the east boundary of Twp 57-22 W4M.)* He said he could just as easily have drawn the boundary another six miles east."

Now that word was getting around about Eric having 100 per cent mineral titles, but prior to Redwater being drilled and discovered, Harvie was being courted by several companies who wanted to make a deal with him.

Principal among these was Amerada,[19] a firm formed by Degolyer & MacNaughton back in 1920, a synthesis of America and Canada; thus, Amerada. Their original purpose was to explore on both sides

19 Taken over by Leon Hess in 1969, just in time to bid on the September 10 Alaska state land sale.

of the border. After Leduc broke, Amerada moved in very aggressively. They acquired crown lands to the east of Redwater and drilled a succession of dry holes. According to Slatter's book, the proposal submitted by Amerada was not favourable, tax-wise, as far as Harvie was concerned. But it did not necessarily mean that he was casting them aside.

Canadian Oil Companies (not yet the parent of Anglo-Canadian) and their brokerage house parent, Nesbitt Thomson, started negotiations in Montreal. There is no indication of how those ended.

Most importantly, in retrospect, Bill Whaley (senior executive with Barnsdall) was in Calgary. Eric recalled Whaley asking "if there was any possibility of making a deal with us for some acreage ... I told Bill that we were considering several deals and, as soon as he and Honolulu (Oil Co.)[20] were ready, I would see what the situation was ... Personally I would like to deal with companies such as Barnsdall and Amerada, as least so far as the operating end is concerned."

Two other contenders, both Canadian and both in the mining brokerage business, were Sandy Richardson and Al Spence. Eric Harvie refers to the former as having access to larger gold companies "...and some of the directors of these gold companies would add strength to our company." The latter suitor was Al Spence, of Mills Spence (brokerage house), "a very responsible firm" (long since disappeared).

By the summer of 1948 Harvie was concentrating on Empire Trust of New York for finance and trying to decide between Amerada and Barnsdall for operational expertise. Empire Trust is said to have held long negotiations with Harvie, but that firm felt that his asking price was too high. When Redwaterwas discovered in August, Heim - one of the key men (the Vice President of Empire Trust) admitted, "This well confirmed the value placed on the (Harvie) mineral rights and leases in the Redwater field." Both parties quickly came to an agreement on Empire Trust's purchase of 20% of the outstanding shares of Western Minerals (owners of the mineral titles) and Western Leaseholds (the lessees of these same titles, non-arm's-length) with an option to buy up 30%.

The Barnsdall group (Barnsdall, Honolulu, Seaboard and Los Nietos) had for some time been negotiating a farm-out. Harvie had set a cash price of $1 million plus a 12½% royalty to Western Minerals,

20 Affiliated with the Matson Steamship Line, serving Honolulu.

but Barnsdall considered this too rich. When Redwater came in - August 30, 1948 - the whole ball game changed. Harvie's advisors were unanimous that his companies were not bound and could repudiate the deal. Harvie declined to withdraw, kept his word and, on January 1, 1949, a formal agreement was executed with Barnsdall. Drilling was to start immediately but did not get under way until the end of March.

The author's July 1992 conversation with Don Harvie revealed somewhat different information. Some time between the dropping of Shell's option and the discovery of oil at Redwater, Seaboard approached Don's father to acquire all of the acreage to the east of that famous line. A deal was set up, but Seaboard had neither put up any money nor signed any documents. When Redwater hit they immediately came to claim all of their acreage. Eric said, "No, you have not signed. What I will do is give you the E/2 of 17, part of the S/2 of 27, and the S/2 of 15 in Twp 56, Rge 21." He also gave them a half interest in other sections, which turned out to be non-productive. He defended his position of keeping 100% of 21 and 31 by stating that they were nearest to the discovery and he should have them.

Robert M. Bischoff recalls his experiences in Redwater.

> In 1948-49, I was District Engineer for Barnsdall Oil Company at Coalinga, California, developing a recently discovered oil field (Guajaral Hills). One afternoon, this large automobile drove up to the field and out stepped Mr. Dunn, President of Barnsdall from Tulsa, and Bill Whaley, California Manager and Vice President - both totally unannounced. After an impromptu briefing they climbed back into the car and Mr. Dunn's parting words were, 'Bob, I think you are going to like Canada'. Ten days later I was at Redwater, hip pocket deep to a tall Indian in mud, setting pipe on the eighth well drilled there.

The team drilled up the E/2 Section 17, then moved across the river.

Bischoff recalls having quite a time getting parts at National Supply. He also recalls the first production: 2000 gallons of 15% acid. The crude was not clean enough, but after a bottle of Scotch to the gauger, that problem was remedied and it was shipped.

About a year later, Barnsdall was bought out by Sunray Oil Company and Seaboard became operator in Alberta for Sunray, Honolulu and Seaboard. "I was asked to stay and was promoted from Division Engineer to General Superintendent. In May of 1951, I was transferred by Seaboard to Casper, Wyoming, as General

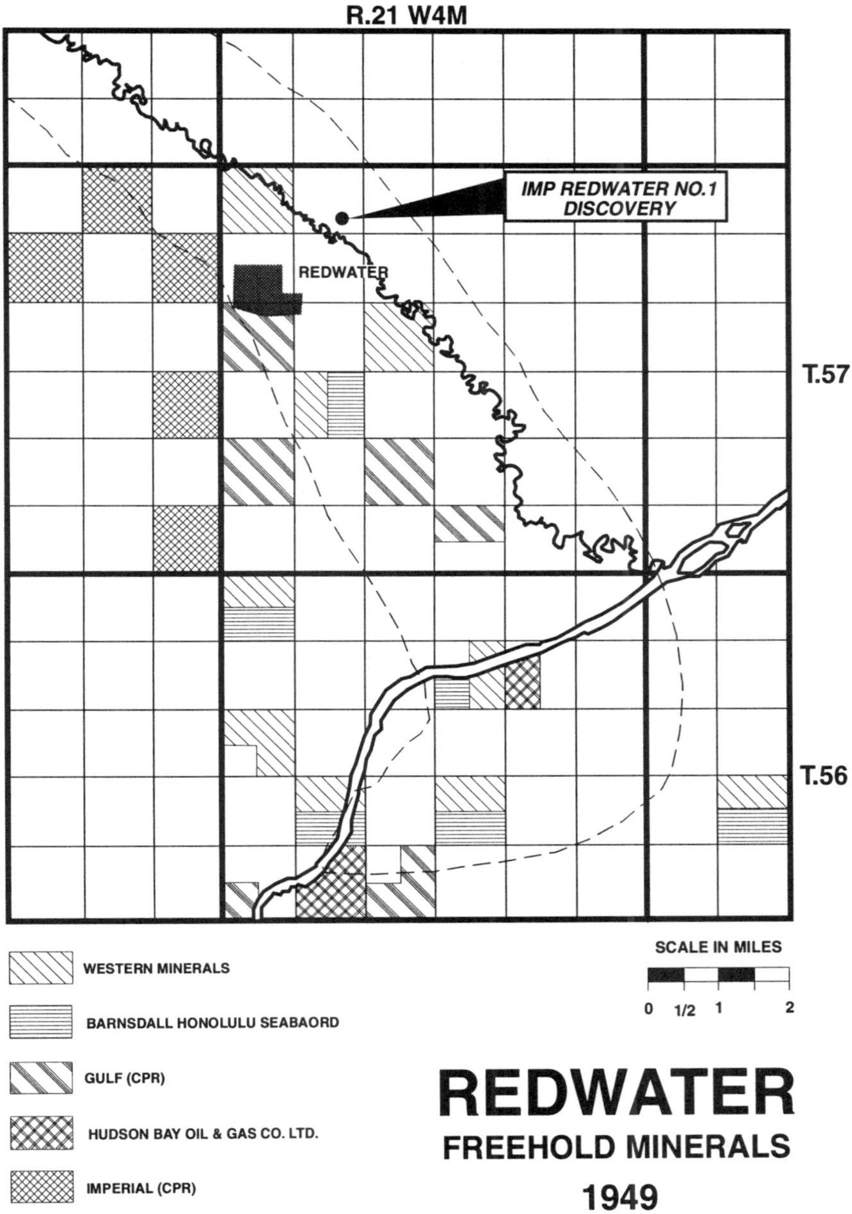

REDWATER
FREEHOLD MINERALS
1949

Superintendent of the Rocky Mountains, and eventually California and Canada." Bischoff went foreign and ended up, just before the Getty crisis, supervising 65,000 employees in 64 countries for Texaco, which by then had takenover Seaboard. He is now retired.

L to R: Bill Walker, Transportation Supt., Bob Bischoff, General Supt.,Dick Borden, Production Supt., Barnsdall hands, on location, E/2 17-57-21 M.
Furnished by H. Bishop

Mr. and Mrs. Joe Budney, owner of property on which 16-17 is located.
Furnished by H. Bishop

John Wasylyk, owner of South Half of E/2(Wells 1-17, 9-17, 7-17).]
Furnished by H. Bishop

Don Harvie claims that prompt drilling of these parcels,[21] resulted in capturing 20% of the oil instead of 10% merely by getting wells on

21 Anglo-Home had started to drill Dec 3, 1948, just three weeks after their purchase of a crown reserve parcel; 16-36-57-22 was on production Jan. 6, 1949. Western completed its first offset well in 12-31-57-21, Feb. 18, 1949. Actually, Anglo-Home really drained some of Eric Harvie's freehold oil!

production before Imperial were able to drill offsets. In other words, Imperial deferred to Western Minerals in getting rigs working on these strategic parcels because they wanted to buy the crude.

Although unsuccessful in cutting a deal with Eric, Dome Mines and Empire Trust were still on good enough terms to ask Western Minerals to help get them into the Redwater oil play. Don Harvie and an officer from Dome Mines in Toronto went up to Edmonton to attend some of the earliest crown reserve sales. They acquired the NW/4 and SE/4 of 31-56-20 and the NE/4 34-56-21. This was before Gallagher was involved with Dome. Western Minerals was to keep 25% of the production and Dome would earn 75% of the production. But, in order to reduce his tax liabilities, Harvie insisted on doing the development drilling, expenditures for which would be tax favourable. Some time after that, because Western Minerals did not have the staff and Dome wanted to continue to be involved, Dome decided to set up an organization and were looking for a manager. This turned out to be Jack Gallagher, then languishing with Imperial Oil's Devon Estates. Many arguments with Hod Meech took place, not surprisingly, over how the tracts should be operated and how the split in revenues should be handled .

Gallagher had been slated to take a senior management course at Harvard. Prior to that the New York office had wanted him to look over Leduc,[22] which he did for four months. In a recent phone onversation, the author asked him to identify stars of those days. He had no difficulty in selecting Doug Layer, Geology, and Jack Harvie, Engineering. Gallagher recalled the order of priority of plays after Leduc: Redwater was No. 3, Willingdon was No. 1. We all know what happened to Willingdon and nearby Pinedale. They were reefs, all right, but they had been breeched by erosion and were full of water. Gallagher subsequently started as Dome's western representative.

Jack Gallagher is reluctant to take any credit for the success of his Crown Reserve purchases in Redwater. He looks upon these as a joint effort with Western Minerals. The one mile-wide corridor which ran along the top of 56-21, and Sec 31 of 56-20 had appeared, according to Jack, as being low, and this is why Imperial did not take those lands when they chose their crown leases. Jack thought otherwise and, with Western, arranged for a seismic contractor to drill deep holes down through the gravel to bedrock. At that point geophones were set on the bedrock and faithfully recorded the subsurface, which

22 The author recalls his first meeting with Gallagher in the Leduc temporary office.

showed it to be higher. The geophones were expendable. These surveys were tied in with short lines along the riverbed. Gallagher recalls that one of the parcels had a pay thickness of 217 ft .

Western Leaseholds/Minerals sold all of their mineral titles to Brascan in 1973. It, in turn, was later taken over by Norcen. Doug Miller, former Brascan CEO, tells how he acquired these titles: The assets included mineral title to 500,000 acres, of which 200,000 were farmed out to Esso on a 999-year lease, and 50,000 to Fina, which later became PetroCan. Much of this acreage was wildcat and was evaluated by Messrs. Baay and Baer. The coal titles were assessed by Chuck Newmarch.

After closing, Eric Harvie and his wife, along with his two sons, were scheduled to lunch at the Petroleum Club. But, because Mrs. Harvie was not permitted to enter the premises, she was sent home with the chauffeur, with the certified cheque for $32 million, and had to have her lunch there.

Imperial Oil's vast CPR holdings, some of which were acquired January 1, 1942 under the terms of Petroleum Reservation (P.R.) #128, ran all the way down to Calgary. This reservation had sliding scale rental fees as follows:

up to 200,000 acres	10¢/acre
next 200,000 acres	5¢/acre
over 400,000 acres	2½¢/acre.

CPR reservation lands could be converted to lease, which then incurred a rental of $1.00/acre and a royalty of 12½% for a 10-year term. Imperial was to "diligently explore"[23] in the general area of the reservation.

In April 1946, CPR acreage was added to P.R.#128 from the following: Twp 56, Rge 22, 23, 24, & 25, Twp 57, Rge 22, 23, 24, 25 & 26, Twp 58, Rges 25 & 26. These lands complemented Crown Reservations 350 and 351 in the same area. Both Crown and CPR butted up against the east boundary of Twp 57, Rge 22. There was some CPR freehold in Twp 57, Rge 21 which Imperial did not request because they had already drawn an arbitrary north-south line marking the east boundary of Twp 57, Rge 22. Gulf would file on it February 6, 1947, to be incorporated into their P.R. #388, which ran all the way down past Big Valley. This acquisition paid off when Redwater was discovered: Gulf moved in Ray Tull's rigs and drilled up 33 successful oil wells.

23 No drilling commitment.

Leslie Munroe, in a masterpiece of bureaucratic understatement: "Through what might be termed an error on the part of Imperial, they did not request reservation of Company rights in the township which later became the scene of the Redwater discovery."[24]

Paul Evans describes Jack Bevel's[25] efforts to obtain CPR freehold and the struggle he had with Gulf (Pittsburgh) to get $25,000 put up. He quoted Bevel as saying, "I'm going to put up the money that's needed. I'll borrow it from the bank. And if Gulf wants it, I'll take it out in their name. And if they don't want it, I'll keep it myself." At the last minute, Gulf coughed up. Bevel could be said to have been responsible for the vast spread of CPR freehold that Gulf acquired.

Gulf must have known that all the crown lands east of Twp 57, Rge 22 - in the heart of Redwater - were open but, at the time of their filing on CPR lands in February, they obviously decided against them. Gulf seemed to prefer fee lands because the rentals were nominal. Besides, they'd had favourable experience with U.S. railroad minerals and it would pay off for them in Canada.

Courtney Cleveland received his university training at U.B.C., then went on to McGill to obtain his PhD. In 1938 he was awarded the CIM's Barlow Memorial Medal for his paper on the Bralorne and Pioneer Gold Mines of British Columbia. During World War II, he prospected the Mercury Belt up through Takla and Pinchi Lakes in that province.

Courtney remembers being in the Vancouver office of Austin C. Taylor, who at the time was controlling Bralorne, along with the V.P. of Production and Exploration, Mel O'Brien, when the mineral title acreage subsequently acquired by Eric Harvie was offered to them. Courtney looked at the plat, but he cannot remember the year in which they examined this spread. At that time, none of the three had much knowledge of oil and gas. Why did Austin, an otherwise wily promoter, turn down this bonanza? Later on his company, ACT Oils, got into oil and gas.

In 1948 Cleveland made several trips to Alberta to investigate the oil business with the idea that he might switch from hard rock to soft rock. In the latter part of March that year, he visited the author in his Leduc office; since then, they have been close friends. He had met several people in Calgary, one of whom was Les Clark, geologist for Barnsdall.

---

24 Internal Report dated April 18, 1959, by Munroe, CPR Manager of Resources.

25 Bevel was "Agent" for Gulf Oil Corp. for Canada (head honcho).

Courtney receiving his 50-year CIM award from Tony Anton, retiring president, May 11, 1993, at the A.G.M. in Calgary.

Furnished by CIM

Cleveland's diary records that he hired on with Barnsdall January 25, 1949, after a hurried phone call. Old Les showed his true colours, giving Courtney only 20 minutes to make up his mind. His first day in the office was February 7th, then Les sent him out to wellsite work north and east of Edmonton. One of these was the first hole drilled by Barnsdall on the Western Minerals (Harvie) acreage at Redwater (16-17-57-21-W4M). Courtney arrived at the wellsite April 18th and reported 32 ft. of porosity in a core which Les ordered him to describe foot by foot to ascertain the pay - no easy task for a person unfamiliar with soft rock lithology. He then went to the south offset in LSD 9-17 and reported coring for green shale on April 30th. Cleveland's diary for May 4th shows they got oil on the first core and reached TD on the 7th.

Ab Lonsdale, sample catcher for Rod Morris, built 65 core boxes for Cleveland, who then visited with Rod on May 2nd presumably to get some information from the Royalite logs (Section 15, which Royalite had bought at the February 15 Crown Reserve Sale). Although Courtney mentions Rod in his diary, he also recalls Reg Shatford, who had worked in the Red Lake Mining Camp (a kindred spirit).

The last entry in Cleveland's diary pertaining to Redwater is dated May 18th, when both Les Clark and Gene Powell (Les's boss) visited the operations. Courtney recalls arriving on the wellsite around ten o'clock at night and looking out over the expanse of the area with rig lights everywhere. Despite Les Clark's domineering ways, Courtney transferred to Pacific Pete with him. Later Cleveland switched back to his old love, hard rock, and managed Pacific's mineral interests.

C. B. ("Boss") Barlow A full head of wavy white hair, an erect bearing, a deliberate manner of speaking succinctly. With his nintieth milestone behind him, all these attributes could belong to a person some forty years younger.

C.B. was born in a Tennessee farmhouse, 90 miles from Nashville. His parents farmed timber from the hill country. He attended rural school until Grade VIII and left home at age 15 to see the world. He rode the train west to Utah "as far as my money would take me." A variety of jobs in that area included work in a copper smelter at Garfield, 15 miles northwest of Salt Lake City.

In 1920, he pushed on to California and started in the oil fields. He had a brother and two uncles, old cable tool hands who were living in that State, so he wasn't entirely alone. He started as a roustabout at $3.50 a day 70 miles west of Fresno at Coalinga, a field containing two grades of crude, 14· API on the west side of the structure and 25· API on the east side. His experience with the heavy crude stood him in good stead when he decided to come to Canada 20 years later.

His first experience as a rotary roughneck was at Santa Fe Springs. He recalled the Bellevue blowout which went on for six weeks and ultimately caught fire after producing 10,000 barrels per day.

C.B. married in 1934 and, when asked where he lived, "Furnished apartments. There was no need to rough it as we would have to here in Canada." They moved five times the first year they were married.

Experts had said, "No oil north of 49 degrees latitude." Neither he nor his peer, de Koch, believed this statement, so they moved to the Vermilion area in 1942. C.B. had an understanding with de Koch[26] that he could give a year's notice and then go out on his own. In 1943 Barlow acquired a 10% working interest in de Koch's interest in what would become Eric Harvie's multi-million-dollar spread, 50¢ an acre for 500,000 acres.

26 Elegant dresser, white gloves.

Taken at unknown wellsite in Vermilion, Alberta, in 1942. Left to right, C. B. (Boss) Barlow, then about 40, Driller (unknown), and William de Koch, Geologist from California who partnered with Barlow. De Koch was always elegantly dressed, with a tie, although his sleeves are rolled up here. The rig is a Star Spudder and in the background is an old Chev pick-up which C.B. had bought from a farmer. Note the high boots and breeches worn by C.B.

Furnished by C. B. Barlow

"Boss" moved to Lloydminster in 1945 and recalls drilling in that area for $4.50 a foot plus day work. They would get by with 15 feet of conductor pipe and, usually, no surface casing. He remembers being shut down because the allowables set by Husky Oil (the only purchaser) cut production back to ten barrels a day. Cannar acreage at Vermilion also produced heavy oil, which was loaded into the tenders and burned directly under the boilers of the steam locomotives on the CNR line (part of the war effort).

On one occasion, when he was travelling to California, Boss stopped off in Calgary at Eric Harvie's office. Some wells had been drilled prior to Leduc, probably in partnership with Eric, at Rusylvia, Alberta. Harvie had already been involved in many areas, acting as counsel for Turner Valley independent companies. He had experience. C.B. states, "He was an oil man and didn't know it" and ultimately "He made all of us quite well off" - the understatement of a lifetime. It is believed that C.B. ended up with a small working interest in the Redwater, and other, producing areas.

When Redwater broke, Eric got in touch with C.B., who agreed to ram-rod the operation. "Many people had tried to lease Eric's properties in the Redwater area (including Imperial) but to no avail." C.B. also comments on the wisdom of having Imperial Oil drill up Western Leaseholds' two sections (21 and 31) on a cost-plus basis (cost plus 10%). Not bad, when one figures the high quality of Barlow's supervision. Imperial had access to tubular goods because a boatload of pipe from Peru had arrived - a result of that company having had to sell International Petroleums back to "Jersey". Western were unable to acquire pipe because of extreme shortages. Imperial needed Eric's crude and provided an instant market for it.

C.B. describes how he managed the project. He started by approaching George Bannantine, then Imperial's Redwater superintendent, up front to hire engineer Ed Campbell away from Imperial. Barlow said that he had also hired Rod Morris only because he knew that Rod was planning to leave. Rod's antagonistic attitude toward the Jesus Departments[27] would limit his stay with Western Minerals; nevertheless, C.B thought he was a top man. And he was.

Barlow brought his production hands from Lloydminster: twins Jack and Lorne McLennon, as well as Roy Hartling and Bill Scutchings. The next job was to line up rigs. This enabled C.B. to drill up Sections 21 and 31, the crown jewels of the field. Western went on to drill 40 more wells in other parts of the field.

C.B. recalls the frantic drilling during the 1949 boom at a luncheon September 23, 1992 at the Ranchmen's Club. "When are you going to get on such and such a location?" Eric would ask ... "We've already drilled it." Meetings in Redwater were, at best, chaotic because of the pace of activity. On one occasion, Empire Trust, New York, congratulated Eric on drilling some more wells. He couldn't

27 Any supervisory person or company.

understand why but his associate, George Crawford, stated "Do you know a nicer way of being embarrassed?"

The next big event in C.B.'s career was the Peel Plateau Eagle Plains exploration play in the Northwest Territories of the 1950s. Tons of 'materiel' were trucked or barged up to drill the first wildcat on this enormous indicated structure (mapped by the Geological Survey). It led the way to further drilling but, unfortunately, with little success except for some very coarse Pennsylvanian sands in one of the wells which yielded some oil.

When asked about his philosophy, C.B. states, "Every day I put in seems like a holiday. I couldn't enjoy anything more. If I had my life to live over, I'd do it in the same manner." Gibby Gibson: "I enjoyed my association with C.B. I've learned a lot from him."

Gordon Hawkins started off as rod-man for Pete Baptie in 1942 when Ivone St. G. Burn was Chief Landman, Imperial Oil Ltd. Gordon was promoted to Landman in 1945.

Gordon Hawkins
Furnished by G. Hawkins

Once Leduc had hit, Morinville was the next target. There had been some geophysical indications of a structure up there. Hawkins recalls assembling freehold leases in preparation for drilling. It would be the first exploratory well to be drilled by Imperial Oil after Leduc in May 1947, but it was a bust except for a few minor shows of gas.

Hawkins states that, as an Imperial subsidiary, Royalite were restricted to that part of Alberta south of Red Deer. Imperial didn't really want to sell Royalite but, rather, wanted 100% control so that they could get to Royalite's cash either in the form of bank accounts or cash flow revenue from Turner Valley. But they had not reckoned with the minority shareholders, who were to be offered a much better price for their stock by Dominion Securities. This led the way to Royalite being acquired by the Bronfmans in February 1949.

Gordon stayed with Imperial, but it was never clear as to which company he was working for. Imperial's office manager, Sid Weller, exercised absolute power over the expense accounts[28] of underlings. Gordon was allowed expenses for only one week while looking for lodging on his move back to Edmonton. Shortly after that, Gordon contacted Rex Dawson (who by now had taken over the Land Department from Ivone Burn) and said he didn't know how to spell 'resignation' but that he knew the word 'quit'. He was loaned to Royalite while they were getting organized earlier in the year and switched over to Royalite permanently in the fall of 1949.

Gordon's first experience at Redwater was going up to get a Surface Lease. The farmer and family were all gathered, with all kinds of booze on the table. Gordon said, after the first drink, "I'm not going to conduct any business. We'll go to Edmonton tomorrow with the lawyer," which they did. Needless to say, Gordon enjoyed himself. One of the people at the meeting said, "You speak Ukrainian?" "No," said Gordon. "You're the first person that ever pronounced my name right, so you must be Ukrainian."

It was Gordon's idea to purchase farms outright where possible and thus avoid wellsite and production facilities annual rental payments. Imperial sent John Evasiw out to obtain options. Each of these options was $100, and they provided for the purchase of a farm at $100/acre, a price better than the going market. All the farmer had to do was pay municipal taxes as rent and continue to farm the land. Some of the farmers started to complain about this and wanted to get out of it. Rex Dawson and Gordon rented a hall in Redwater and called a meeting of all those so affected. Rex explained Imperial's position and offered to tear up any of the options, if they wanted, and they could keep the $100. Not one came forward, which quieted that issue.

Hawkins moved up through several companies, ending his career as Land Manager with PanCanadian, CPR's oil and gas subsidiary with control over millions of acres of freehold in 1989. His extensive knowledge of, and access to, CPR records has provided much valuable help to the telling of this story.

---

28 Weller forbade the author to stay at the Palliser, the York Hotel being good enough for an underpaid geologist.

This chapter closes with the unbelievable story *of how a world class oil company ended up empty-handed in Redwater while its arch-rival emerged with two thirds of this reservoir. The facts herein disclosed for the first time provide an accurate step-by-step account of Shell's calamitous "strategy."*

The Dingman discovery at Turner Valley in the summer of 1914 alerted Imperial Oil to the possibilities for hydrocarbons in other parts of western Canada. Section 40[29] of the federal regulations (Alberta still did not have its minerals) forbade non-Canadian companies from filing on Crown lands, and this included Jersey's subsidiary, Imperial Oil. They were, therefore, restricted to CPR freehold.

In 1917 Shell Transport Company Limited of London, England - Jersey's arch-rival - wanted it all. They asked Ottawa for all those lands north of Twp 69 in Alberta[30] plus a portion of the Northwest Territories along the Mackenzie River totalling over 300,000 square miles or 200 million acres. Shell did not fall under the Section 40 stipulations because it was British-owned and, consequently, not deemed to be foreign. It was Shell's contention that scientific development by a single company would eliminate the stock frauds, or swindles, that commonly attended North American oil booms.

When Imperial Oil learned of this incredible land grab, they issued grave warnings about the "yellow peril."[31] Alberta, not yet owning its resources, chimed in with objections of its own. Lobbying must have paid off, because Shell retreated - not to return until 1941 - opening offices[32] in Calgary under a different guise with direction coming from Los Angeles.[33]

The Alberta and B.C. Foothills were Shell's main targets, preceded by surface geology, with many of the early geologists serving on field parties. It is not certain when geophysical activity started, but the first Foothills wildcat, Shell Norman, was spudded in July 1942 and abandoned at 12,000 ft in October 1943. It was after this that Shell very likely had Cec Chesher re-work the seismic records. The next

29 Breen, David H., "Alberta's Petroleum Industry and the Conservation Board," University of Alberta Press, 1993, pp 18-20.

30 This included Shell's present Peace River extraction facilities plus all of the Keg River plays (Rainbow, Zama).

31 Shell had already begun to use yellow paint on their service stations.

32 Shell Oil Co. of B.C. Ltd., #721, 300 - 8 Ave S.W. (Lancaster Bldg.) By 1944, Shell Oil of Canada Ltd: 116 - 8 Ave S.W; in 1946: 809 - 4 St S.W.

33 Alex Clark (later to work for Home Oil) was transferred to Calgary from L.A. in 1943. Les Clark (no relation) was still in Centralia, Illinois.

hole was started in May 1944, and it discovered the Jumping Pound field with an open flow of 17.5 MMcf/d and 4200 psi bottomhole pressure. Three more step-outs had been drilled by 1947.

As has already been noted, Shell signed a $30,000 option in May 1946 to tie up hundreds of thousands of acres of checkerboard freehold on the following terms: *mineral titles*, not leases, with a royalty of only 2½% to Eric Harvie. This spread of land included millions of barrels of undiscovered oil under Sections 21 and 31 of Twp 57-21 W4M in the heart of what would be the Redwater reservoir. If Shell had heeded Chesher's report of the anomaly on its seismic line over Redwater, it may have been a different story and they need not have dropped their option. And, if they had realized there was open crown land, they could have filed on that prior to April 1947. By then, Imperial filed on Crown Reservation #433 - 200,000 acres of previously undisposed Alberta Government land covering the as-yet undiscovered reservoir. There was no bonus requirement, merely filing fees.

Shell dropped their option late in 1946 despite the fact that two wildcats were drilling in the Leduc area. By that time (1945) Chesher had left. (Cec, now heading up Canadian Exploration Company, shot Leduc with his own crew before road ban in the spring of 1947 and confirmed the feature.) But Shell's decision to pull out of Alberta had been made by The Hague.

Carl Nickle's Daily Oil Bulletin of October 19, 1948, provides the official excuse for Shell's 1947 decision: "Shell is going to end 'sideline' sitting in Alberta and re-enter the active play." The announcement came from London, England, where Shell Caribbean Petroleum Company of New Jersey was arranging for funding.

The article states that

> ...Shell Oil Company of Canada started exploration in Alberta in 1941 and by 1946 $7 million had been spent, most of it on Jumping Pound. Shell had been hard hit by WWII, and Alberta was one of the areas selected for slow-down. A few months after that decision was reached, the Leduc discovery was made...
>
> ...Shell's large exploration department in Alberta began the break-up; however, in line with the 1946 decision, some went to New Brunswick and others went to other American units of Shell...
>
> ...Shell *did* have money for the northern Alberta foothills, participating with four other major companies in "Operation Muskeg," a $2 million project...

> ...Among the deals dropped by Shell was an agreement covering extensive freehold rights[34] in central Alberta. That proved a 'heartbreaker'. Some of these rights are now producing oil in the Leduc field. Others are sitting in close to the new Redwater discovery. These are the breaks of the oil game. Shell played an important role in the pioneering days of Alberta foothills and plains exploration preceding Leduc. Conditions brought on by the late war forced the group to retrench - a tough decision - made doubly tough by the subsequent limelighting of Alberta...

> ...Now the Shell group has arranged for a quarter billion dollars capital. It is a logical decision and a welcome one, that the group plans to put a slice of these funds into Alberta.

And all of this at the time when Shell's Alberta staff consisted only of geologist Ian Crawford and a secretary.

Several reliable witnesses have reported on the unbelievable anomaly which turned out to be Redwater:

(1) Les Clark, who went with Seaboard, told Jack Browning in 1949 that the feature had shown up on seismic;

(2) Gordon Hess reported that Brookshire, party chief for a Heiland crew (under contract to Shell), had shot to near the top of the structure and, despite wanting to continue, was told to stop;

(3) Peter Savage inspected Shell data in the Robinson Estates (basement of Penny Lane, 8 Ave & 5 St SW): old seismic cross sections taken from seismic markers showing unmistakable structure;

(4) Chesher, himself, who decided to quit Shell after they had ignored his reporting on the Redwater structure, dismissing it as a regional feature without merit. Chesher told D. P. McLaws, Q.C.(after he had shot Redwater and had seen the structure) that The Hague had cabled the fateful order to get out. Marsh Porter's identical statement to the author back in the '50s corroborates this eviction order;

(5) Wilf Baillie[35] states that Shell had run a single reflection correlation line all the way from Spruce Grove (west of Edmonton) using highways then crossing Redwater in 1945. It is said to have shown up a large plateau with high ridges. (Was this line planned or seen by Chesher?);

---

34 Shell dropped their option on Eric Harvie's minerals at the end of 1946.

35 Baillie joined Shell in the summer of 1945 and continued a successful career with them.

(6) Wes Rabey recalls Les Clark telling him that a Heiland crew under contract to Shell had recognized east dip.

Shell were ordered to move to Ottawa. The Company worked there and then proceeded to New Brunswick, where prospects were said to be much brighter, being along the coast with salt beds and possible domes plus ready access to an eastern U.S. market

According to Fred Kidd, when Shell were pulling out they were said to have offered the Jumping Pound sour gas field to Imperial Oil for $6 million, but by that time Imperial had discovered Leduc, so they had no further use for supplies for their now-scrapped synthetic gas plant. Besides, they were already strapped for cash because of Leduc drilling commitments.

Ian Crawford graduated in 1944 and was hired by Les Clark, of Shell, in the same year. He points to Nickle's October 1948 article in defence of Shell's managerial expertise. The altogether-too-short Eric Harvie option period saw Shell drill Fribourg, west of Lloydminster, but it was located on crown lands. Shell's only previous plains venture occurred in 1942 at Castor (Shell Chinook). Both were dry holes.

Jack Browning gives Les Clark's version of Shell's operations in Redwater area. Les had been a senior geologist with Shell who went to Seaboard and, later, Pacific Pete. Les said Shell had <u>all</u> the acreage, but this is patently impossible: C.R. #443 (200,000 acres of virgin crown lands) had been issued to Imperial Oil in June 1947, and no one had had title to it prior to that date.

Ed Fulmer, then with Chevron, was bird-dogging the seismic crew which confirmed the reef after Imperial Redwater No.1 was discovered. The main purpose of his crew was to get a handle on the Crown Reserve parcels that were being put up - in particular, the February 15th sections.

Fulmer does not know what took place after he sent the data from Edmonton to Calgary for John O. Galloway's perusal. There is no record of Cal Standard having bid. (Unsuccessful bidders are not publicized.) However, Ed knows that Cec Chesher persuaded Galloway to purchase his seismic map of Redwater for $10,000. This was based on Galloway's trust in Cec from Southern Alberta days. Ed reckons that Cec probably made all his money back in three sales and much more after hooking up with Frank McMahon.

Ed also mentioned that Cec would save money by going into other contractors' shot holes rather than drilling them himself.

Peter Bediz, distinguished retired geophysicist, remembers Shell's modus operandi after their return to Alberta in 1949. They had insisted on a degreed person to supervise Century's two crews slated to move up to Canada, and Peter was chosen "...just for three months, while they get started." He never returned to the U.S.

Bediz thinks that Shell may have misinterpreted Redwater originally because, he said, they were overly cautious in trying to correct for the glacial drift to perfection! This same concern was equally shared by other geophysicists. He said they often questioned: "Did you correct properly for weathering?" This was a "must" in normal geophysical prospecting. He describes the firm as being "Always theoretically motivated, highly technical, didn't want to take chances[36] (the European way)." They, like many other companies, told Peter, "We want another Leduc."[37]

Bediz made it plain that the field crews weren't supposed to.know too much. The procedure was to carry only one reflector horizon, and that for confirmation, to avoid "busts." But, as Peter recalls, busts *did* occur. Companies didn't want contractors to carry any horizons, but he insisted that his parties do so. That is the mark of a true geophysicist and a purist, a seeker after the truth.

36 Ian Crawford insists his company did take chances.

37 In 1962, Shell took over Canadian Oil Companies, successor to Anglo-Canadian. One of Anglo's many assets was a one-half working interest in three productive quarter sections in Redwater; thus, Shell finally did obtain a minor position in Redwater - but nowhere near the prize they could have had.

7 THE LAND: CROWN

A Battle of Wits: Imperial Oil vs. the Crown

Ever since September 1, 1905, the Alberta Government had been trying to lift the dead hand of federal bureaucracy to allow the province rightful access to its resources. A statute related to oil and gas administration had been set up in the late twenties by provincial legislators who were applying pressure on Ottawa, but it was never made law.

Alberta was finally granted administration of all its resources in October 1930, but the enabling legislation was not passed until June 1931. Hubert Somerville, then a newly-appointed clerk, was given the task of sorting out one of the four express train carloads of documents from Ottawa and transferring the federal records into a new provincial filing system that he had devised. In 1937, new regulations limited oil and gas reservation areas to 50,000 acres each, but they carried no drilling commitment. This is a holdover from the old mining days when all one had to do was walk over the terrain looking for ore bodies.

Hubert Somerville at the rig raising, Devon wellsite, August 1990.
Furnished by Author

Hon. E. C. Manning, Premier of Alberta 1943-68, at his son's tea party, June 18, 1993.
Furnished by Author

By 1941, the promoters (many of them bankrupt) had given up except for minor roles in Turner Valley and Wainwright. In the absence of any serious exploration, not to mention significant discoveries,

Alberta now looked to the majors[1] for action. The Government, therefore, took steps to attract those companies accustomed to tropical concessions (South America and the Middle East) by allowing them to file on as many as three tracts of 200,000 acres each. Lease sizes were increased to a maximum 9600 acres (15 sq mi).

When Leduc hit, the administration could see that it had to change gears quickly to protect the Provincial Government's crown lands which had now become valuable. On May 26, 1947, Ivone St. G. Burn, Imperial's head landman, applied to Hubert Somerville - by then superintendent of Mining Lands - to convert all of Imperial's crown lands in Twp 50, Rge 26 (the discovery township) to lease. Not any too soon, as Ivone warned Gordon Hawkins, his assistant, on his return to Imperial Oil's Calgary office, "No more 100% lease selections". In July of that year, maximum reservation sizes were reduced to 100,000 acres (no more than two to a customer). This was easily circumvented by the formation of numerous dummy subsidiaries and the use of employees' names as figurehead reservation holders. These nominees, of course, had no equity, nor would they ever have. Lease blocks were (incredibly) increased to a maximum of 16 sections. Crown Reserve legislation, on the books since 1941, was quickly wheeled out. Lease selection by a reservation holder triggered the creation of an equal amount of acreage in the form of Crown Reserve in each township. These latter tracts would be disposed of by the Crown as it saw fit.

When Imperial hit oil at Woodbend (Twp 51, Rge 26, all crown) immediately north of Leduc in January 1948, they were entitled to apply for a 16-section lease block which just happened to cover almost all[2] of the productive D-3 reef. Edmonton had legally given away the farm.

Hubert Somerville and Hon. N. E. Tanner[3] lost no time in rectifying the situation. They called industry to Edmonton on March 1, 1948, and advised that the maximum size of leases would be reduced to 2x4s (two miles by four miles) and 3x3s (three miles by three miles). Industry complained this was retroactive, but the Government reminded them that provisions for crown reserve had

1 Gordon R. Hess, retired geophysicist, claims that Shell were assigned the area west of Hinton and east of Vegreville, and Imperial were allotted the Edmonton area - neither being formalized by the issuance of Reservations. Hess recalls instructions to ship barrels of Lloydminster crude directly to The Hague.

2 See pp 150-151, LEDUC (S.A. Kerr, 1988).

3 Minister of Lands and Mines (Mines and Minerals after April 1, 1949).

been on the books since 1941. On March 29, 1948 an Order in Council legalized these new lease dimensions. Thus, when Redwater hit in the late summer of 1948, Edmonton was ready to play hardball on Reservation #443. This almost 200,000-acre spread had been filed on by Imperial in April 1947, after the D-2 was known, but *before* the D-3 had been encountered at Imperial Leduc No.2 on May 7th.

Sometime between the March 1st meeting and the March 29th deadline, Imperial did an end run by converting its Crown Reservation #350, which ran as far as the east boundary of Twp 57, Rge 22, into 4x4 lease blocks to maximize its acreage position. The easternmost of these leases created a one mile-wide crown reserve buffer to separate it from #443. In so doing, acreage for the first crown reserve sales in Redwater was immediately ready for posting shortly after the discovery. These first tracts started the ball rolling for a rapid succession of sales.[5] Did the crown respond to industry's request for posting, or did they set up a few parcels as trial balloons? We will never know, although Hubert Somerville claimed that only those lands requested by industry would be considered for posting.

Following the discovery of oil at Redwater No.1, Imperial chose a completion date to give them as much lead time as possible to select a lease under the new rules. Although oil had flowed on drillstem test August 31st, the date set for "the presence of oil in commercial quantity in the opinion of the Minister" was fixed at October 1. Neil Alexander McDonald, titular holder of P&NG Reservation #443, was notified by letter under Somerville's signature as superintendent of Mining Lands that he had three months to apply for a lease or leases containing the discovery well. He was also reminded that no wells could be drilled within 4½ miles of the discovery until the application for the lease had been made. Imperial diligently hewed to this distance requirement and moved both Imperial Redwater No.4 (later known as Egremont in the northwest) and Imperial Redwater No.3 (later to be known as Imperial Amelia to the southeast) to locations which would meet these stipulations. It is noted that the original Application for Reservation (under 1941 rules) carried with it only "...any investigation relating to subsurface geology and geological and/or geophysical examination in conformity with the regulations..." with

4 This was Hubert's idea. He also remembers industry's attempts to try for 8¾ square miles.

5 See Appendix XVIII for list of Crown Reserve parcels (1948-52) furnished by H. H. Somerville.

no provision for obligatory drilling. But Imperial knew it would have to "drill or drop" and their seismic pointed the way.

As of January 1, 1949 maximum reservation size was to be reduced to 100,000 acres. This meant that Imperial had to split the old #443 in two. They, therefore, requested an extension on unleased portions of #443 and the newly created #443A. At the same time, McDonald assigned "his" reservation to a conveniently incorporated subsidiary of Imperial known as Constellation Petroleum Limited.

Concurrent with this was the submission of an internal progress report to Imperial's office in Edmonton: "An 18-man seismograph party conducted refraction shooting for a total of eight days; 26 profiles were obtained from 16 shot points at an estimated cost of $5,000. In addition, a six-man gravity meter crew worked two days to establish 67 stations at an estimated cost of $450."

When the smoke had cleared away, drilling credits of over $130,000 had been accumulated, along with other items, to practically balance the rentals paid by Imperial on all the leases up to May 1950. So Imperial got all of its rentals back. Even using Ken Jeffrey's solemn declaration of May 19, 1950: $138,605.24, the finding cost is micro-fractions of a cent per barrel. Why would Imperial want to bid on any of its erstwhile reservation lands, which would only up the cost of crude to them, when it was already entitled to 100% lease blocks gratis?

Imperial Redwater No.3, the southeast step-out, was completed as an oil well on January 2, 1949. This created another lease selection requirement, but Imperial did not reckon with Hubert Somerville when they submitted their first choice, which included 2½ sections in the north part of Twp 56, Rge 21. Hubert's letter to Imperial turned down this request but pointed out an option which would create a corridor. "There is no objection to your changing this lease concentration to comprise a four-mile by two-mile block cornering to the southeast of Section 22-57-21 W4." This parcel actually was reduced to 1½ by 3 miles with one section and two half-sections tacked on (S/2, all of 3, E/4 and all of sec. 10 in 57-21), creating a pseudo 3 by 3-mile lease block which would meet Hubert's cornering provisions. Sections 34, 35, 36 in Twp 56, Rge 21, and Section 31 in Twp 56, Rge 20 thus became crown reserve and yielded the highest bonus prices per acre of any of the posted crown lands. Hubert's insistence on maximizing the crown's position thus put many extra millions into the provincial coffers.

The first sale of Redwater Crown Reserve parcels was announced by John Harvie, Deputy Minister, October 19, 1948, less than three weeks after the official completion date of the discovery well. It took place November 10, 1948. These were straight-up leases with no override but with a May 31, 1949 drilling commitment. Parcel No.1 was the NE/12-57-22, ridiculously distant as it turned out to be, four miles to the southwest of the discovery. Bids on it were rejected, and it was never re-offered .

The SW/36 was posted in four 40-acre parcels, but all bids were rejected. Had this been arranged to appease the CCF[6] by making one-well locations available to the down-trodden "little man"? No more LSD-size tracts would ever be offered in Redwater. Anglo-Home stunned the community by bidding over $2,000/acre for the NE/36, but people should have realized they were buying oil in the ground at so much a barrel. This bonus set the pace for future sales.

James C. Scott had conducted important geological field work in the Alberta foothills for Shell Oil, and then was transferred to eastern Canada when that firm was ordered out. Jim returned to Calgary and joined Husky Oil, then owned by the Neilson family of Cody, Wyoming. Their main thrust had been at Lloydminster because of their experience in heavy oil. It was, therefore, somewhat of a departure for Husky to consider Redwater crown reserve acreage.

Cec Chesher was on contract with Husky, promoting his entrepreneurial skills by agreeing to split the bonus costs of any crown reserve purchases with Husky in return for a working interest.

Jim's analysis of the NE/36-57-22 W4M, which was posted for the November 10, 1948 crown reserve sale, showed it to have considerable merit. Scott recalls the meeting at which Neilson Sr. was shown Jim's results and his estimate for a bid, assuming about 10¢ a barrel in the ground. Alex Bailey, former ERCB Chairman was, at that time, landman for Husky. Neilson stated that Jim's proposed bonus was too rich for their blood, so Husky did not submit a bid. This was the first, and the last, time that they would look at Redwater. Jim knows that his estimate would have beaten Anglo-Canadian's successful $351,000 bonus.

The next sale, December 17th, re-offered two of the above parcels, but bids were once again rejected. So, at the end of the year, Anglo-Home ended up with the E/36 while Pacific had bought into several tracts with partners.

---

6 Commonwealth Cooperative Federation, later to become NDP.

C. O. Nickle, in his Daily Oil Bulletin, criticized Edmonton for rejecting tracts. Was it because he or his partners had been turned down?

> The government has rejected bids on some of the Redwater parcels. That action is quite within the powers of the government. A lot of oilmen feel, however, that by taking that action the government is setting a bad precedent. The industry is probably better qualified than the government to judge the value of oil rights, especially those that are 'unproven'.
>
> Redwater rights in a less favorable structural position than the four tracts which sold for over a million dollars total are hardly worth as much, at least at this stage. In fact, some of the rights still unsold will probably wind up valueless.
>
> It seems to us that the government would be wise to accept the oil industry's judgement on values of Crown Reserve tracts, just as it has apparently accepted the industry's judgement on 'royalty bonus'.
>
> Another point raised by oilmen is the method of offering Crown Reserve rights, using the 'sealed tender' method. Bulk of the industry, we believe, is in favor of open, competitive bidding - with each bidder, and the amount of his bid, being known to the industry and the public. In lieu of open bidding, it is recommended that the names of 'sealed tender' bidders, and the amounts of their bids - whether accepted or not - be made public.

On November 25, 1948, Imperial spotted Redwater No.5 on CPR freehold minerals lands (LSD 9-25-57-22) only two miles west and south of the discovery. They could drill with impunity because limitations on proximity to discovery applied only to wells on crown land. It turned out to be a dry hole, abandoned because of its back reef location (poor, water-bearing porosity).. However, this hole, along with further seismic, enabled Imperial to more intelligently select their first lease block.

To recap the first sale, the E/36-57-22 was purchased by Anglo and Home for about $650,000. They moved in rigs immediately, and the net result was that the wells were on production one month before the first Western Minerals hole to the east had been drilled. Indeed, Gord Connell recalls that Anglo-Home were producing 2000 bpd per well[7]. This meant that Anglo and Home were draining Eric Harvie's offsetting oil inn Sec. 31. The law of capture had never been repealed.

7 Paul Greenwood, ERCB, said 26 trucks hauled crude 24 hours a day.

Hastily passed Alberta legislation forbade anyone from suing either the Crown or Pacific Petroluems for any negligence at the notorious 1948 "Atlantic No.3" blowout,. The end result of this was $1 million[8] being handed over to Frank McMahon by the Conservation Board in January of 1949. Pacific Petroleums' (partners included Max Bell's Calvan Consolidated, Garnier, and others) bids for the NE/34 and the NW/36 were most likely made possible by this windfall.

Cec Chesher (whose company initials, CEC, stood for Canadian Exploration Company) was shooting the field for Pacific[9] in return for a 1% override and 10% working interest on every parcel that Pacific bought. *McMahon is said to have bought Cec's interpretation, cutting down on the chance of competition.*

The second sale, which took place December 17, 1948 listed only two quarter sections, both of which had been previously rejected. Pacific Petroleums bid on NE/24, putting up $127,000 for what turned out to be a "bust".

The number of McMahon's partners varied; for instance, Sunray came in for 25% and a few months later sold out for a handsome mark-up. They reappeared later. But one thing Frank excelled at - sucking in enough partners with sufficient funds to ensure he ended up with a piece of the action at *no cost* to him.

Partners in purchases, incredibly, totalled more than 100% according to documents filed in the PetroCan system:

Pacific Petroleums	25%	
Sun-Ray Oil	25%	
Princess Petroleum	25%	
Canadian Exploration Co	10%	(Cec Chesher)
Calvan Petroleums	23%	(c/o Max Bell, The Albertan)
Total	108%	

Later in the year - October 27, 1949, to be precise - Sun-Ray wanted out and sold their interest in two quarters for $450,000, having received their share of production up to that time.

Leader Pete (ownership unknown) acquired a 75% interest in LSD 9, Sec. 25-57-22, a farm-out through Princess Petroleums (one of McMahon's companies) September 28, 1949. This well, which had

8 Monies held in trust by the Board and deemed payable to Pacific Pete (McMahon) after all costs had been recovered .

9 Gordon Hess, by now Chesher's partner, moved his seismic party out to Redwater immediately after the discovery and shot every edge crown parcel that was posted, taking a percentage of production (if any) for their data.

absolutely no hope of being a producer, was drilled in December 1951 and was subsequently abandoned.

By 1958, McMahon's labyrinthine arrangements were becoming even more complicated. In addition to Sun-Ray/Calvan, the following partners with widely varying interests appear in the PetroCan files:

Garrison	Goliath
Polaris	Admiral
Zodiac	New British Dominion
Godfrey	Charter
Anaconda	Leader Pete
Garnet	

The Notice of Crown Reserve parcels (by now proven acreage) posted January 12th asked for "offers to purchase petroleum and natural gas leases up to the close of office, February 15, 1949". They consisted of Sections 15 and 30 *by quarters* (Parcels 16-23).

> "Anyone may ask that if his offer on one parcel is not accepted that he be considered on another parcel providing his order of choice is clearly stated and the amount of his remittance is sufficient to satisfy the purchase price, rental and fee on the further choice."

Immediately after the announcement, some members of industry made representations to Hon. N. E. Tanner that a bidder might be encouraged to offer a higher bonus if he had the opportunity to bid for a lease of the entire section and thus reduce his drilling and operating costs. Mr. Tanner bowed to this pressure and directed that a new notice be issued to include a key paragraph reading:

> "Inasmuch as only offers and not tenders are requested, a person may include in one offer all of Section 15 or all of Section 30 to be considered in relation to the total of the offers being the highest for each of the four parcels comprising such section."

Carl Nickle, in his February 5th Daily Oil Bulletin responded with a quip, "Wanta buy a whole section?"

John Harvie, Deputy Minister,[10] let it be known that he disagreed completely with the amended proposal (to consider offers for the sections as a whole) because it might eliminate bidders whose budget could afford only one quarter section.

10 This would be his last sale. He was transferred to "Lands and Forests" on April 1, 1949, and Ian McKinnon became Deputy Minister of the new "Mines and Minerals".

British American had been trying for over a year to obtain production by either farming-in or purchasing parcels at Woodbend. Their glimmer of hope had been Frank "Sugar" Schultz's Cretaceous discovery over in the northeast part of Woodbend. B-A were desperate, paying out the astounding sum of $450,212 on December 12, 1948, for Parcel 15 which would ultimately yield only two marginal wells on the north extremities of the tract. On January 20th, President W. K. Whiteford announced a big B-A program. Ducking the question of why in hell B-A hadn't found any oil on its own, "Independent producers have always supplied us with most of our needs." At that time they were paying bonuses of 10¢ to 15¢ a barrel over the posted prices.

The creation of the Redwater nine-section block in the northwest part of Twp 57, Rge 21 following the initial discovery begat Sec. 30 as crown reserve. British American, this time with Hudson's Bay Oil & Gas as partners, were determined to buy their way into production. And so the February 15th sale provided the opportunity; the team purchased all four quarters in Section 30 for an aggregate of $2,044,000. But they had not realized the leases were on town property,[11] and this upped their drilling costs. Furthermore, the SW/30 proved to be too far southwest into the back reef to be productive.

From 1920 until early 1949, Royalite had been a majority-owned subsidiary of Imperial Oil which, in turn, was 70% owned by "Jersey". But New York head office had decreed that Imperial had to fund their own development activities. They were, thus, forced to dispose of Royalite. Minority shareholders had received better offers from other financial houses, forcing the Bronfman Brothers, through Dominion Securities, to improve the purchase price for Royalite[12].

Gordon Connell had been a long-time Royalite employee, and when they split off on their own, he stayed with them. He was involved in putting together the prospectus to sell the properties to Dominion Securities using Ford, Bacon and Davis as engineering consultants with the help of Bill Gallup, then chief geologist. Dominion Securities, in turn, sold the Bronfmans' equity into CEMP - the acronym for their children's Trust. Royalite were given a healthy dowry, over $6 million, and were ready to spend it as fast as they could.

---

11 When B-A woke to the fact, they rushed up to Edmonton to plead for reduction in the bonus. Although Ian McKinnon was sympathetic, Hubert Somerville curtly dismissed the delegation, "You should have known better."

12 See pp 35-38, "Corridors of Time".

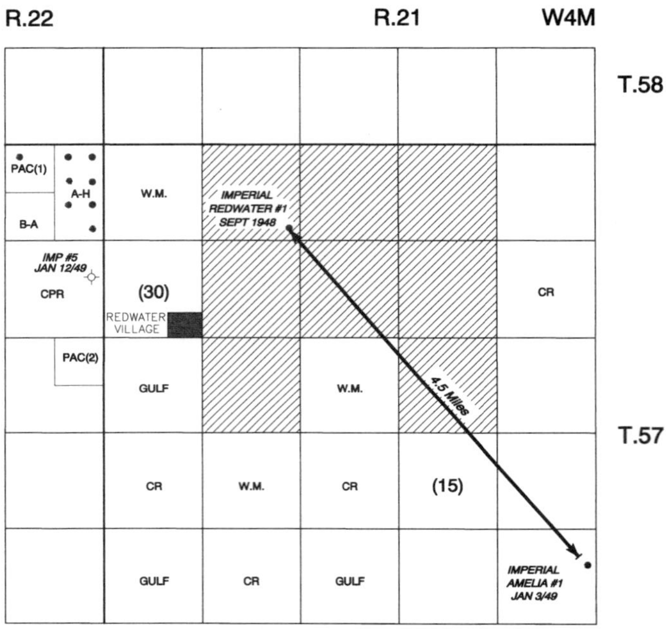

REDWATER CROWN RESERVE SALES
NOVEMBER 1948 - FEBRUARY 1949
SHOWING BONUSES PAID TO THE ALBERTA GOVERNMENT

(hatched)	IMPERIAL OIL CROWN LEASE
GULF	GULF CPR FREEHOLD
W.M.	WESTERN MINERALS FREEHOLD (100% TITLE)
C.R.	CROWN RESERVE
PAC(1)	PACIFIC PETE et al - NOV 10/48 - BONUS: $101 000
A-H	ANGLO-CANADIAN & HOME OIL - NOV 10/48 - $653 000
B-A	BRITISH AMERICAN - DEC 17/48 - $450 000
PAC(2)	PACIFIC PETE et al - DEC 17/48 - $127 000
(15)	ROYALITE - FEB 15/49 - $3 153 491
(30)	BRITISH AMERICAN AND HUDSON'S BAY OIL & GAS FEB 15/49 - $2 044 580

Gordon is the only survivor of those present when Royalite were deciding on the size of tender they would submit on Section 15. "A bunch of us sat in the Board Room (606 - 2 St. S.W.) and worked over the bids. We discarded the high and the low, and we took the average of the other ones." The following people are said to have been present:

S. F. (Bobby) Heard, President
King Houston, General Manager
Chick Banister, Purchasing Agent
Everett Chambers, Legal Counsel
Fred McKinnon, Exploration Manager
Gordon Connell, Chief Engineer.

On the basis of 100 feet of pay, using the parameters later developed by Irene Haskett, some 200 barrels per acre-foot recoverable, *Royalite would be buying oil in the ground*, assuming 55% recovery factor, for 25¢-33¢/barrel.

When asked if he had obtained any information from Imperial, specifically Imperial Redwater No.3 in 9-11-57-21 W4 which had just been completed (January 3), Gordon remembers Don Mackenzie telling him that the well was making about 1,200 barrels per day but gave no other details.

On the day of the sale, Royalite submitted a tender for all of Section 15: $3,154,171. Royalite's letter stated that, if the above bid were unsuccessful, the monies be apportioned as follows:

Parcel 17	NW/15	$1,226,171
Parcel 18	SE/15	1,051,171
Parcel 16	NE/15	869,171
TOTAL		$ 3,146,513

The letter was silent as to Parcel 19, SW/15 simply because Royalite knew that the sum of these skid bids would use up most of the available cash and have insufficient left over for another tender.

The Minister and his Deputy were in attendance when bids were opened. The Crown accepted Royalite's $3,154,171 for all of Sec. 15, despite Harvie's plea that this bid should have been ignored. He wanted Royalite's tenders on 17, 18, and 16, plus the tender from an unnamed bidder on Parcel 19, to be accepted. When one compares Royalite's bid for 17, 18, and 16 totalling $3,146,513 against $3,154,171, there is only a $7,658 difference. We will never be able to prove it, but this difference was far below what another company would have been reasonably expected to offer, if for no other reason than that it held *proven* reserves. Assuming the tender on Parcel 19 had been $500,000, the Crown can be assumed to have forfeited ($500,000 - $7,658) $492,342 by not accepting individual bids on each of the four quarters.

One must also conjecture at what would have happened if another company had out-bid Royalite for the entire section. This would have released Royalite's funds to compete on the three quarter sections, plus whatever may have been tendered by someone else for the fourth quarter. Regardless, the award on Section 15 was sanctified with the passage of O/C 204/49 February 23, 1949 citing "...by advertisement dated January 12 *and as amended*, published in newspapers within the province..."

How much hard information was really available from Imperial Redwater No. 3 (Amelia No.1) to (a) Royalite and (b) the companies? It was supposed to be kept tight. How much data did Imperial slip under the table to Royalite? How much weight did other bidders give to their seismic information? There is a lot of Low Velocity Layer along the valley floor running through Section 15. This could give false lows in the hands of inexperienced geophysicists and lead them to believe that the reef was at a lower subsea elevation because of the slow velocity through the LVL shallow beds. This, in turn, would result in less apparent pay.

Amerada may have submitted bids on the Royalite parcel for no other reason than that they had been very active in the area and were also one of the contestants for Eric Harvie's freehold. They must have shot the structure geophysically and probably saw what Imperial had seen. But, to repeat, had they misread the LVL?

One final footnote from the "Royalite Oil Reporter":

> On February 15, 1949, Royalite purchased 640 acres of Crown Reserve, covering Section 15 at Redwater, paying the Alberta Government a cash bonus of $3,153,491. Later, a 12½% interest was sold to a partner, the Company retaining 87½%. By the middle of June, 1949, 16 crude oil wells and one gas well were completed on this tract. At December 2, crude oil sales had amounted to 438,231 barrels with a gross revenue of $1,202,510.88. After payment of normal Crown royalty and operating costs, seven-eighths of the net revenue accrued to Royalite.

For the one-section sale of June 15, 1949, did the Minister wish to test the waters again (for the last time, as it turned out)? He posted Sec.16, not only on a quarter-by-quarter basis but subject to offers on the whole section:

> "Inasmuch as only offers and not tenders are requested, a person may include in one offer all of Section 16 to be considered in relation to the total of the offers being the highest for each of the four parcels comprising such section."

The sale results were shown as one number for parcels 31-34: $3,223,320 under the signature of A. T. Jacobson, Amerada's[13] Calgary manager.[14] Gordon Connell comments on the above purchase. "We wanted to bid $3,600,000 but the Bronfmans said it would require at least $4 million, and they concluded, 'Don't bother'." The sale results greatly upset Gordon and Fred McKinnon because they could have had it if they had insisted on their $3.6 million bid. Based on their Section 15, which they estimated at 18 million barrels recoverable, the oil would have cost them 15¢-16¢ a barrel in the ground.

Fred Kidd tells a story of Shell upon their return from eastern Canada. Their attention was called to Sec 16 of Twp 57 Rge 21 W4 by an expert who is said to have assured Shell that it was worth $4 million and would be an excellent purchase. Obviously, Shell did not act on this advice.

Somerville states that an individual had submitted a bid on one of the quarters but, overall, Amerada's aggregate bid on all four quarters exceeded those of other contestants. Hubert, now acting Deputy Minister (the Department of Mines and Minerals was created on April 1, 1949), ruled that Amerada's bid be accepted. This eliminated the need to explain publicly how the bid on one 160-acre tract was higher than one quarter of the $3,223,320 but was rejected.

A Sale That (Almost) Soured

Bids containing certified cheques or cash were to be submitted "at the close of office" (5:00 p.m.) in Room 403 of the Administration Building (109th Street, Edmonton). All the offers were spread out and checked to ensure the highest bidders. Then, at about 8:00 p.m. the side door on the ground level would open and interested parties would enter to hear the results. Hubert Somerville recounts the story of a tender that went missing (for a few hours).

Ivone St. G. Burn (who had left Imperial Oil in 1948) had submitted a bid on a Crown Reserve parcel on behalf of Scurry Oils Limited. But when the successful bidders were announced, Ivone's tender wasn't called out. In accordance with Section XVII of the Act, Somerville was representing the Minister who had the right to review decisions of staff members. Ivone told Hubert, "I think our bid was higher." Hubert immediately checked with his staff, who admitted

---

13 Amerada: American & Canada, incorporated in 1920. Hess, A U.S. marketing firm, took over the company in the summer of 1969.

14 Offices then in Blow Bldg, 212 - 7 Ave SW.

that the Burn offer was on the last page of the Scurry letter which had been overlooked. He examined the letter, confirmed "You're right," and made the announcement in favour of Ivone.

The now-unsuccessful bidder cried out, "Now everybody knows how much I bid!" Hubert merely replied, with a wink and his inscrutable smile, "And they also know you had more than $100,000."

Gordon Hawkins frequently visited Edmonton to attend the Crown Reserve sales, along with many of his U.S. colleagues who had come up from the States. They had very mixed feelings about the sealed bid because tenders were always announced after each U.S. sale. They thought Government officials could be switching numbers around and committing other forms of dishonesty. They soon learned that Hubert Somerville was scrupulously honest and would never countenance skullduggery in the back room. Nor would he have any time for any land man whom he caught trying to make an "end run" around him. Their consensus was that the Government had cleverly designed this system to extract the maximum bids, but they admitted that it saved the winner a lot of embarrassment as to how much he had left on the table. Only the losers knew, and they would rarely tell.

By April 29, 1949 information on the Imperial Oil step-outs was common knowledge. All the Crown reserve would have been set up by that time. Although desperate attempts were being made to get holes drilled, road bans were in effect and it was virtually impossible to move rigs. Out of seven parcels posted on that date, only three were purchased - all by Pacific Petroleums and all in the northwest extremity of the field. Pacific must have still been feeling bullish despite the fact they were on the back reef. Nevertheless, they put their wells on production at the full allowable.

The July 15, 1949 sale saw 13 parcels up for grabs, bids on five of which were rejected. These tracts were on the east edge of the reef. A new player entered the game: Texaco Exploration Company. Texaco had yet to discover Wizard Lake and Bonnie Glen; these finds would have to wait until 1951. Along with B-A they, too, were desperate for crude and purchased four quarters, only one tract being fully productive. The northeast quarter of Section 10 fetched $626,000 from Anglo Canadian. Once again, Pacific bought two quarters for $100,000 each.

Increased industry awareness, coupled with the promise (not always kept) of being able to buy oil in the ground, was creating unheard-of activity at the sales. Calgary landmen with cheques in their hands were a common sight as they alighted from CPR trains at the Jasper Avenue station on their way to submit tenders to the Department of Mines and Minerals on 109th Street.

There was only about a six-week gap until August 25th. "East edge reef" was purchased by Royalite and by British American separately. This was one parcel in which B-A finally scored, but had to pay over $900,000 for the quarter, setting a new record of $5,600 an acre. A newcomer, Ed-Bridge, bought a dud - SE/3 - for only $157 an acre. This sale marked one of the few times that 15% overrides were tacked on by the Crown. This presumably enabled Calvan (with Pacific as partners) to purchase two quarters on the east edge. On the minus side (B-A was always on the minus side), B-A bought an edge parcel on the west side with the override. Little production ensued.

Alberta Giant (N. E. Tanner's son-in-law?) purchased a dud on the east side, NW/8-58-21 for $206/acre and a 15% ORR.

The next sale was October 27, 1949, at which Parcels 62 to 70 were posted. And three new players came on board, one of which was Ohio Oil Company with the purchase of an east reef edge tract; Canadian Gulf also bought a tract even though they already had a freehold (CPR) position.

Local financial entrepreneurs, the Tanner brothers (Harold and Reg) had assisted in the purchase of Royalite from Imperial through Dominion Securities and were looking for opportunities. They acquired two choice quarters in the northwest, resulting in the name of one of the companies being Lease 64 Limited. British Petroleum came to Canada in 1953 and bought out these and other tracts in Redwater. They used Tanner's name, Triad, then switched back to BP. In 1992, Talisman acquired all of BP's assets and in 1993 sold off Redwater production.

The next sale was not until February 16, 1950, at which Parcels 71 to 82 were offered for the first time south of the North Saskatchewan River, southeast of Amelia. By that time the Imperial-Simmons step-out, completed on October 1, 1949, had resulted in creation of lease block. Royalite submitted the successful bid: $9,494.36 an acre for Parcel 77, the SW/36-56-21. Assuming a 55% recovery, and about 27,500 bbl per acre, this would work out to 18¢/barrel. You would, therefore, bid $5,000 an acre. And if we look at the same parameters, Royalite's bid on the NW and SW/36 (Parcels 76 and 77, respectively, fetched bids of about $9000/acre) works out to about 35¢/bbl in the ground. These tracts were underlain by the very porous seaward type rock reworked by waves yielding breccia type which graded into massive limestones and muds, calcaranites and fossil types of rocks.

R. A. Brown, Jr. formed Redwater Petroleums Limited shortly after the 1948 discovery. As with many promoters, his company name was a misnomer designed to lure unsuspecting shareholders into his lair, as he had no funds with which to buy crown reserve. On the advice of Ted Link he filed on over 500,000 acres of crown land north and east of the field and made a deal whereby Barnsdall, Seaboard and Honolulu took farmouts. Apart from a few gas wells, this venture ended in failure. Brown then changed the company name to Time Oils.

His fortunes were soon to change with the purchase of Imperial Oil's subsidiary, Foothills (the Bank of Commerce put up $2.65 million for him, in those days a virtually unheard-of loan). Revenue from these Turner Valley wells now gave him sufficient money to consider purchasing crown reserve parcels in Redwater.

Brown organized a syndicate to be known as Highland Leaseholds: Brown's company, Federated Petroleums - $1.28 million; Wood Gundy - $900,000; General Petroleums, New Superior Oils, and Devonian Limited - $1.31 million; United Oils (Bobby's other company) - $180,000. This enabled them to purchase on May 11, 1950, Parcel #87 (SE36-52-21 W4) for $1.6 million, the then-record, plus $1.59 million for the adjoining quarter section (#86, NE36). Brown then brought in Leduc West Oil, Spooner Oils, Scurry Petroleums Ltd. (these latter two through Spi Langston), Jupiter Oils Ltd., Calta Assets (Brown family's private company) and Cody Spencer, head of General Pete. This latter consortium was called Riverside Leaseholds Ltd. On June 29 they purchased an adjoining (#95, SE35) quarter for $1.85 million. Royal Trust was the "front" bidder for all three tracts.

Arranging for the drilling of these three quarters had been ramrodded to some extent by John Scrymgeour and Bill Atkinson, Brown's key henchmen, who - in order to get the earliest maximum return - had lined up the farmer's land the day before the sale and had a rig ready to move in next morning. There was other financing (Prudential Life Insurance Co. of New York), and all of the net production revenue was hypothecated to them according to the author's recollection. Maybe they made application beforehand on the assumption they would acquire it, but the Board would have needed evidence that they *were* the beneficial lessees.

When Bobby took over Home Oil from Major James R. Lowery in 1951 he hired R. B. Curran from Imperial Oil, but "Rapid Robert" turned out to be a disappointment - spending more time at the Petroleum Club, the horse races or World Series baseball than he did in the office. Upon Curran's abrupt departure Brown switched the three parcels into Home Oil's portfolio on a share exchange basis. They would stay there until Triad bought all of Home's interests March 15, 1961 for $7 million. John Hrynchuk, Anglo's field superintendent, states the abrupt physical takeover was New Year's Day, 1961.

Industry realized the importance of reservoir studies - keeping in mind Irene Haskett's dictum: "The type of rock material present is the controlling factor." Royalite continued to buy parcels (SE/1-57-21 for $6,930 an acre). Another newcomer, C. L. McMahon Inc. (not to be confused with Frank or George), purchased a tract in the northwest extremity along with Bishop Oil.

The final crown reserve sales took place in 1952: the N/9, N/10, and N/11, purchased by Bay Pete,[15] Gulf and Admiral/TransEmpire.

Jack Browning is an RCAF veteran, a graduate of the University of Saskatchewan, and after many corporate vicissitudes is now on the board of a thriving local oil company. His time is interspersed with foreign tours and the publishing of his late wife's family reminiscences.

At the time of Redwater, Jack was Pacific's chief geologist reporting to Lorne Falconer, field manager operating out of Edmonton (97th Street loft). He had obtained his job through Art Nauss, then with Tropical Oil (a Jersey subsidiary). Art had returned to Canada and wrote Jack offering him a position. Pacific did Bear Oil's[16] work, which covered hundreds of thousands of crown reservation acres north to Ft. McMurray. Link, also temporarily connected with Bear Oil, thought that some Leduc oil had migrated and was trapped in that area.

Pacific had little funds of its own despite McMahon's unjustified windfall from the Atlantic No. 3 disaster. Browning said Frank would farm-out pieces of every crown reserve parcel he bought, charging enough to result in Pacific retaining a healthy working interest at no cost to them. Browning states that he worked out the geology of the

15 Not to be confused with Hudson's Bay Oil and Gas Company Limited. Bay had a strong U.S. integrated presence; sold out in 1955 to Tennessee Gas Transmission on the impending demise of Mr. Bay, the owner. Jack Browning, who took over at Bay, had little use for these properties because they were on the south limits of the reef. Browning later went with Tenneco.

16 Bear Oil was owned by four companies but had no staff.

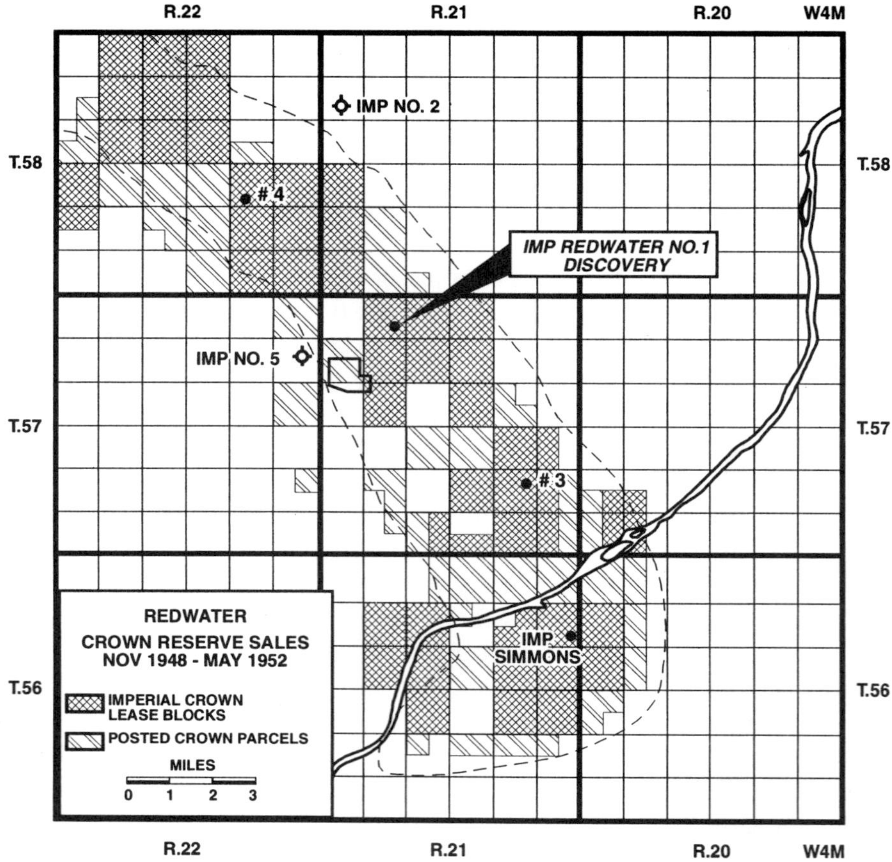

Plat showing all Crown Reserve parcels posted between 1948-52.

Redwater reservoir and did all the calculations for bids on the posted parcels. "All bids were modest, relatively speaking, and were successes."

Les Clark, geologist with Seaboard Oil - which was going to be shut down - needed a new job. According to Browning, Clark exerted pressure on Cyrus Wright, Sun-Ray's head and part owner of Bear - also affiliated with Pacific. Cyrus is said to have told Frank McMahon he wanted Les to have an important job. McMahon obliged, mainly because of Sun-Ray's position as one of Pacific's partners and a front for the U.S. pipeline, Pacific Northwest, which he was hoping to supply from his northeast B.C. gas fields. When Browning learned of this appointment, he quit 45 minutes after Les arrived; 32 other geologists left in 13 months. "It was like revolving doors."

Browning had another brief fling with Redwater when he took over Bay Petroleum Corporation's Canada operations. This consisted of three crown reserve tracts on the south extremity of the pool which had been purchased before he got there. "They were just junk."

There were stories that the RCMP escorted roughnecks on to leases to drill wells because the farmers were stopping them. Jack O'Brien's first job in the oil patch was as a wellsite geologist for Pacific Pete under the tender tutelage of Jack Browning. He recalls the fights in the beer parlours between the farmers and the roughnecks, and whenever he went in for a beer he made sure he sat against the wall.

8 THE PEOPLE

Sodbusters vs. Roughnecks

Ed Lakusta, grandson of immigrants, was born on his father's farm near Opal, Alberta in 1930. He has lived through the profound changes from an agrarian economy to the oil boom, of which he became a key player. Ed, now watching his "home place" from a distance through the lens of his brother's elk farm, sums up the unique nature of Redwater's community, a pot-pourri of ethnic backgrounds:

> ...I think, if you look at the mixture of people ... there were all nationalities. I grew up with ... displaced people, French, Japanese, Scots, English, Germans ... We have a very heterogeneous community, but predominantly Ukrainian ... They came out in the early 1900s and were in the majority...

Religion was a driving force and a foundation in the lives of most of these immigrants and, more often than not, the topic of conversation when they met. This was especially true of the Ukrainian community. They would spend hours discussing profound issues surrounding Catholicism and oppression in the homeland caused by religious strife and how their families[1] had withstood and survived their troubled history. They do to this day - a people intent on preserving their traditions and beliefs.[2]

Although the Lethbridge area in Alberta is known for its well established Japanese community, few realize that Redwater had its own settlement of Japanese families.[3] The Redwater area provided a refuge for them, where they were no longer harassed or displaced. Their Buddhist faith and spiritual ties with the past had carried them through many dark days and helped them to maintain composure throughout the frantic boom years.

Prior to the discovery the village of Redwater contained only 100 souls, all with close agrarian ties. Ukrainian holidays were observed, as was the Sabbath. John Hawrelko, author, retired educationist and champion of the Ukrainian esprit recalls his father being forced to go out to his fields on a Sunday to salvage some of his crop. John's aunt berated his father unmercifully for his transgression - breaking the Sabbath. Yet, a few years later, one couldn't tell night from day; the whole world turned around when the 1948 boom hit Redwater. Exigencies of time and money now meant 24 hours a day,

---

1 Some of which are profiled in the Blue Pages.

2 See Appendix XIX "A Brief Historical Note" by Peter Yachimec, and "Ukrainian Beliefs" by the Author.

3 The Kimuras and Kiyookas are two prominent families profiled in the Blue Pages.

L to R, Virginia Sharek (Malowany), Mary Malowany,[4] Marge Wright (Malowany) and Grace Malowany (Peter's wife) at the Wright residence in Calgary discussing ecclesiastical matters.

Furnished by Author

seven days a week - materialism had infiltrated the stern mores of the local inhabitants.

In the words of those who were there...

John Myroon, retired farmer:

...When the No.1 well blew in, we actually felt bewildered. Our lives were changed forever, and we did not realize it. We thought the oil people would soon leave and we'd return to our quiet pastoral life. But that was not to be. From a small community which was mostly Slavics, the farmers who constituted at least 75%, became a minority. Our language, our lifestyle and our culture disappeared into a confusing, multi-ethnic whirlpool. Helping one another was replaced by greed. Many social ills befell the community. Some became rich while others were left morally impoverished...

4 Profiled in Blue Pages.

Obutsudan (altar) of the Jodo Shinshu branch of the Buddhist faith in the Kimuras' back room.

Furnished by Author

John and Olga Myroon with aerial photo of their Redwater farm.
Furnished by Author

Anne Key (nee Yachimec), teacher, mother and historian:

...It was just terrible ... people slept on the sidewalks ... people worked all night ... we'd never heard of working on Sunday ... and the noise, and the roads ... they tore them up and they sank down in them ... and they blamed us!

Walter Hrynchuk, part-time farmer and part-time oil worker:

...People lived in chicken coops , granaries, anything that put a roof over their heads. They parked anywhere it was possible. There was no facilities such as running water or sewers.. or electricity, even.

Jim Evasiw, postmaster:

The rigs were coming in, and the people were just flocking in. By the end of '50, there was well over 3000 people, and there was no accommodation! We had a little shack at the back. We used it as kind of a hothouse and also a storage. You know that a fellow came in.. I had to almost literally throw him out. He said, "I am going to live here.. I got nowhere to stay."

The Brown house, a typical pioneer home with log siding and thatched roof, became a haven for migrant oil-workers. At one time, there were nine families living in the yard in tents and trailers. The singing teakettle on the Brown stove never quit its tune. The Pasemco farm outbuildings also accommodated many a worker.

The Brown House (NE/20)

Furnished by Author

There was no "room at the inn," so to speak. Many settled on farms and in little hamlets surrounding Redwater, the town of Opal being one. It was one of the biggest towns around, with a full-sized railway station, five farm implement dealers, a hotel, two or three elevators, post office, two blacksmiths and four stores. John Hunchak, never one to miss an opportunity, purchased a quantity of war surplus double decker bunks which he put up in the basement of his hotel in Egremont.

Jack McCaskill, the first petroleum engineer on the scene, failed to find lodging in Redwater. He finally located a shack at Radway owned by a railway section foreman (a part-time farmer who had lost his wife). The McCaskills paid what was then considered an exorbitant rent: $40. On top of this the shack was impossible to heat; Jack's electric bill was $40 a month even though they were using an old oil stove.

Based on previous experience at Leduc and Devon, Imperial Oil recognized the need to establish a "company town." They scouted surrounding communities for a possible townsite with Egremont being the favoured location mainly because it was on the railroad track, and there was room for expansion. However, they were unable to get Adolphe Purschke, a well known Egremont land owner, to agree on a price. They eventually settled on a site adjacent to the original Redwater settlement.

Gibby Gibson describes the townsite in detail:

> ...They purchased a bunch of land on the north side of Redwater and laid out a town site ... a residential area. There was no thought of putting up any businesses or things like that. And within a couple of weeks they'd moved in an office-bunkhouse accommodation which was the old army H-hut type. Two wings were bunkhouse, and the other two were for the offices. I was in one of the wings. I guess there would be 20 of us in one, beds side by side and a little stove in the middle. A little gentleman used to come in and fill it up with oil once or twice during night, depending on how cold it was. And they had toilet facilities installed. But the water! You couldn't drink it! So we drank bottled water brought out from Edmonton. That's where we spent the winter of '49. But while all this was going on, they were busy constructing some proper bunkhouses, two-to-a-room type with a lounge in the middle ... a very adequate field office for us, and warehouse, pipe racks ... We also put up a smaller building to accommodate three or four girls that were working in the office for us...
>
> They showed a lot of foresight when they laid out the town. They only built on every second lot. There was a marked sameness in a lot of the houses. Their thinking at that time - and this has certainly been the case as other people move into the community - they will build in between and have different style homes, so you won't get the sameness. And the cost of building them will simply be a tie to the utilities. They also drilled a couple of Viking gas wells ... By the early summer, we had all the amenities of home: gas, light, water, sewer...

Jack Harvie was responsible for construction and adds his own perspective starting with orders from his manager, Tip Moroney:

> ...In the spring (1949) Tip called me, "Look, we need a bunk house for 60 men up there." So, I drove up to Redwater. On the way through Edmonton, I stopped at a contractor named Alton & Anderson who had done work for us. And I sat down with Ken Alton and we decided to go with two 30-man bunk houses - in case of fire, you see...

> Calgary did the work on retaining Bird Construction, and they made, sort of, the layout of the town. So we took those plans, and Tommy Osborne, the manager of Bird Construction, supervised the building of these houses. And Ken Oakley did most of the supervision of the contractor. And after the sewer and utilities ... Ralph Flanders -looked after the pump station from the river and that pipe line...
>
> We gathered the sewage and we ran a line down to the creek - about a mile away - and we put in a great, big filter bed there. We just had an outfall spilling and it went through the filter bed, cleaned up, and then worked its way down to the creek...

According to Pearl Court, wife of a British-American production hand, B-A had promised housing comparable to Imperial Oil's town plans. But when Pearl and her brood moved up from Coutts, in November 1949, they found themselves cheated and disillusioned. Fortunately, husband Ted had built a 18' x 24' shack, which they located on a 3⅓-acre tract owned by Frank Roshko just east of the town. Their first purchase was a stove, which they put in the middle of the one-room shack. There were no facilities. Millions of dollars had been spent by B-A on crown reserve sale bonuses but nothing on the employees. Water line and sewage pipe ran through Roshko's land but they were not to be tapped by the occupants.

But Pearl was no quitter. She recalls her days in the B-A camp (Skunk Hollow) with nostalgia. The winter of 1949-50 was one of the coldest on record: -40°F for weeks. Because there was no electricity, Ted bought Pearl a gasoline-powered washing machine. "Every day was wash day ... Because I was the (surrogate) mother, my young neighbours thought I was also the doctor, comforting and curing ... Solid, enduring friendships were built up from those pioneering days."

Although Imperial's housing was considerably better, contending with the winter of 1949-50 was a challenge for the residents. Kay Jeffery, whose husband, Bernard, was an accountant for Imperial, remembers it vividly:

> ...It was -48°F in Edmonton that morning. The movers were two hours late.. they couldn't get their trucks to go any sooner. And about three people registered that it was 62 below.. in Redwater. And most times, in cold weather ... Redwater was about 15 degrees colder than Edmonton ... And by the time the furniture was in, the temperature in the house was 48 above ... We found the furnaces were inadequate ... never could get the houses over about 66 ... running full blast. So, they had to change the furnaces ... put in more BTUs...

> The weather tested us well that first winter. It stayed quite cold for almost six weeks. Then we were wading in mud for about the next six weeks. To top it off, the soil there was powder-fine, and the wind blew for about the next six weeks. Our houses had shingle siding, and the dust just sifted through. One day after we had vacuumed our dining room, which was on the northwest corner of the house, we checked it after a couple of hours and could write in the dust on the table ... We were all in it together, so we managed to laugh about it sometimes...

Kay summarizes the years:

> ...We had a good life with Imperial Oil. We were still in the era when loyalty was quite strong and sort of a lifelong thing that you had. We made eight moves.. weren't always that enjoyable. But we met a lot of wonderful people. You'd move away from them, and you'd move back to where your different friends were ... The quickest way to get transferred in Imperial Oil was to start building a house somewhere ... when you put all the curtains up. You always said it was good to leave one curtain not done...

Bernard sums up his career.

> ...When I think back over ... 41 years.. I felt they'd been good ... Imperial Oil was an excellent company to work for. They looked after us when we were in the Service, made up the difference between our Service pay and what we were getting with the company when we left to join up. I felt they were a great bunch of people to work with and I ... enjoyed the years...

Office workers were also victims of that winter. Lorraine Fox, geological secretary in the Imperial's "chicken coop" office, had to wear her boots - as did Alice Lukes at work in her husband's drugstore.

The roads left a lot to be desired, and the movement of heavy oil equipment over the dirt surface didn't help since they weren't oiled to start with. Paul Greenwood, Conservation Board engineer, provided a very good description of the various routes to Edmonton, their link with the outside world.

> ...There wasn't a paved road within miles. So, every Friday night or Saturday morning, when you were planning a trip to town, you inquired around as to best available route. There were four selections. The first one was the Old Victoria Trail which, provided it didn't rain, was generally the smoothest. Access to it was six miles straight south of Redwater. The Victoria Trail originally ran from Edmonton to the Victoria settlement on the N. Sask. R. south of the town of Smoky

Lorraine Fox awaiting another geological report.

Furnished by S. Bleakley

Lake. I don't think you can travel it now - I think it's all been cut up by fences and fields and whatnot, but it was a pretty trip and the one most often used...

Another option was to go south of Redwater and cross on the Bruderheim ferry, taking a dirt road until you got to the Ft. Saskatchewan highway, and then on into town. Highway 28, ostensibly the main route, was generally a terrible piece of gravel road all the way to Namao, where you hit pavement for the eight-mile run from Namao Corner in to Edmonton. What a relief!

And the final option was to take Highway 28 to the Gibbons Junction and then go south through Gibbons until you hit the Ft. Saskatchewan highway just across the river from Ft. Saskatchewan, and on in to Edmonton on a piece of pavement through Horse Hills and Oliver. Anyway, it was always an exciting trip, winter and summer and caused a lot of laughs and a lot of broken springs...

John Hunchak laughed at the plight of inexperienced truck drivers from the U.S., believed to be working for Brinkerhoff Drilling. He called them "coons." When the ground was wet and they left the gravel highway to head out on to a side road, they would go no more than several hundred yards before ending up in the ditch, much to the amusement of John and his friends.

Line-ups were the order of the day in the bars, liquor store, post office and bank. John Hunchak recalls the long queues at 10:00 p.m., when still-thirsty patrons wanted to buy their case of beer from the Egremont Hotel (10¢ a glass and $3 a case). There were "square fights" as Alex Chaba put it. He recounts the story of a crowd of oil people and farmers standing in line at the liquor store when someone commented, "We'd move a lot faster if there weren't so many Ukes holding up the line," to which the farmer in front of him turned and told him to take back that remark. Settlement and retraction were resolved by a fist fight.

The Evasiws, who ran the post office, described their situation:

> ...You opened up in the morning and there was a line-up of about 30 people. So you looked up and there was about ten Smiths in there, so you grabbed the S's and you throw the Smiths out, and they sorted it themselves. They passed it on to each other ... We used to work anywhere between 12 to 16 hours. I would do a bit ... and then come home in evening and help her out - until it got to the stage that ... It got so that I told her, you know, that's the end. That's as far as it goes! So, this around about 10 o'clock. And the Superintendent was Mr. Neil. So I phoned him. I told him, "Do whatever you want. If you are not here tomorrow, we are not opening the post office. It is impossible. We cannot handle it." You know that he was here at 7 o'clock the next morning and thought he would give me a hand? Well, in eleven days, our (new) post office was up...

Jim and Anne Evasiw, left, sorting mail in the tiny Post Office.
Courtesy of Imperial Oil Ltd.

There were even line-ups for the telephones. Gibby describes the switchboard in Boston's store:

> ...The town just wasn't geared to accept the influx of a lot of drilling people.... There were two lines from Redwater to Edmonton! They had a couple of telephone booths there, and so it was a pretty hectic place in the mornings as the drilling contractors, the supply people, and other contractors were phoning in their reports...

Rita Astle (nee Cook) describes her experiences in Boston's store, where the only telephone was located. Rita remembers her involvement with the primitive switchboard in Boston's Store. She mentions Eve Royer (later Forry) and herself as Operator III. "We worked at a hectic pace, as we only had one line to Edmonton, and a lot of the oil men would be lined up waiting for me to get their calls through."

The Royal Bank went through its series of crises, too. "They didn't have a place to start banking, so they set it up in our garage" said Mary Malowany.

The bank opened in April 1949, and Norma Torguson (nee Anderson) became their ledger-keeper in May. "Mr. Smith was the manager, and Stan Hughes was the teller."

Says Norma of her experiences,

> ...I didn't know anything about banking ... cheques to be cashed. Some of them, we were stuck with, and some of them were okay... We used to run out of money sometimes. And they had what they called mutilated funds to be sent to Bank of Canada to be replaced with new bills ... like, any old money put away ... We'd sometimes have to bring that out...
>
> We did run out of money ... and we made a trip early one morning, to Edmonton, so we could open up.. The money parcel wouldn't always come out when it was supposed to by daily train. And coins would come in a great, big bag, and you'd pick them up at the train station, but the bills themselves were in a bundle and they'd come through the post office ... by registered mail. And then ... we'd go over to see Jim Evasiw ... And there'd have to be two of us - one was usually the manager, and sign for it - two signatures. And the manager would always take a gun ... We always had a gun in the cage when we were tellers! Every morning we brought out our money and the gun...

Asked about target practice, she replied, "I did once ... just out in back of the bank. You were allowed to fire it once in a while."

The Royal Bank used to make them work Saturdays. Norma (ledgers) and Stan Hughes (teller) in temporary quarters in Malowany's Garage. Note calendar date, vault, and temporary plywood counter.

Furnished by N. Torguson

The education system also had to bear the brunt of the boom. Theresa Scragg, who taught school at that time, remembers having to handle Grades IV, V and VI, a class of 52, in her chicken coop. It resembled a zoo, she said; one-third were oil patch people.

The locals hadn't seen anything like this in their lives. It is not surprising that they were bewildered, confused, and often suspicious. They had no understanding of what was going on, and the oil companies were secretive and non-communicative. Walter Hrynchuk had something to say in this regard:

> ...There was a lot of seismograph crews around this area ... They did, almost constantly for two straight years, all the drilling and testing on the roadways or wherever they could get into ... people's property and stuff like that. They had their own crews, and they just stuck together ... there was no way you could talk to them or find anything out. It was a guessing game. You did not know what they were up to or what they discovered...

And so did John Myroon:

> ...At night, when all was still, we could hear the roar of the machinery, clanging of pipes and yelling of the derrick men. It was exciting, yet scary. Our two boys, who were 7 and 5 at the time - and other neighbour's children, would walk to the wellsite across the fields to look at what was happening. However, they were afraid of the noise, machinery and men, so they would peek in awe from the fenceline or bushes ... Even our animals, which included horses, cattle, hogs, and poultry, were baffled and spooked by the excitement and noise...

Anne Key considered the new arrivals to be "mainly very nice" but then there were those who "came to drink, dance, and came for money ... a lot of prostitutes ... they had gambling - at Peg Leg Pete's -just across from Malowany's Garage." Mike Senych, retired educator, observed hostilities toward newcomers, who were perceived to have taken over the place. He recalls the sociological shock. The oil workers worked hard and played hard. And there was disdain for the oil wives who smoked.

Attitudes varied on both sides - rubes and roughnecks:

Evelyn Gavel, the wife of the mechanic on Lineham's crew, arrived before the boom: "They were lovely people ... they went out of their way for us ... we weren't intruders." One day she was asked, "Your name, Evelyn, is Ukrainian ... Are you Ukrainian?" and the Gavel children were picking up the language from their new playmates. Later on, Fin Lineham wanted to have a Christmas party but could find no place to hold it. After some negotiation, because some deemed it would be sacrilegious, the townspeople reluctantly agreed to let the crew members hold their function in the Ukrainian Hall. And, too, their Christmas comes two weeks later.

Gordon Astle, roughneck who later married Rita Cook, remembers Redwater as a typical small Alberta town consisting of a large percentage of Ukrainians. "We found the area very friendly, and they immediately included us in all local functions."

There was little for the newcomers to do except to eat and drink. There were three main sources for alcohol: the beer parlours, liquor store and Yarmula, the bootlegger. As for the latter, John Myroon referred to him as "A dangerous man ... After the war, we had a celebration ... and he invited us to his yard. I could never drink moonshine, so I drank beer. He knocked that bottle out of my hand. He says, 'You think you're better than us? You better drink this stuff!'"

On another occasion Bill Allen needed permission for a lease of acreage for a battery site from Yarmula. By that time they knew each other pretty well, so the first thing that had to be observed was the visit to the manure pile in the back of the barn. After prodding through the ordure, Pete retrieved a bottle. He poured a drink for Bill and said, "Drink." And Bill stated that when you are told by a Ukrainian to have a drink, you drank. What an experience - because Yarmula was an acknowledged leader in moonshine booze in the Redwater area.

Being confronted by hundreds of hungry men who had put in tough shifts calls for quick solutions. Hugh MacKenzie talks about the Mess Deck...

> ...Well, hardly any of the guys cooked their own ... went over to the old Mess Deck café ... First we went to the old Quonset hut that the hotel had put up as a temporary eating place after it burned down. And the one thing you ate there, the eggs, were not too bad. But I ... remember

Manning Drilling Company rig on a Gulf Shkolny hole. Note U.S.-style housing. "Mess Deck," on extreme left of picture, fed as many as 1500 men daily.
Furnished by H. MacKenzie

they had a deaf and dumb cook and a Chinese cook there, and them two would get into an argument in the morning sometimes. And the deaf and dumb guy would be just squealing, and the Chinaman would be swinging a knife.. And you ate your meal, and you didn't sit around afterwards to have a cigarette with your coffee or anything, because there was about four guys waiting for your seat...

Gibson comments on how the locals and the roughnecks coped with the fire which destroyed the old Redwater Hotel.

...But they were really a fine bunch of people there. They rolled with the punches. I guess I could give you one example. The hotel, just shortly before I arrived there, burnt down, and with it went probably the only eating place in town. The people that owned the hotel - within a very, very short time - had acquired one of these old army huts, 20 or 30 feet wide and 60 feet long, and they set up a restaurant in there. And real service! You didn't have much of a selection, but there was lots of it, and wholesome. I remember, a little later on, I was working in the office, and I went over for a sandwich - which was all I felt like for lunch. I got there just before the big rush for the noon meal and they asked me if I wouldn't mind coming back later in that there were people who were hungry coming in. (laughter) So I said, "No, I think that's fair enough"...

The Redwater Hotel fire in late 1948 left memories with quite a few people. Denis Blakeman, native of Black Diamond, describes Redwater.

...Just another typical little town ... had a dirt street ... The hotel was on one side and a block away was a little café ... another of those many little towns we had been through. It had a railroad and elevators ... a hotel which later burned down, and a few stores ... You didn't really associate with the townspeople ... You might go to the local dances...

...One of the problems we would run into is that when we would arrive on the drillsite, every driller and tool-push thought that, because you represented Halliburton ... you were a brother ... I can remember people complaining ... they got up to pay their meal ticket and the waitress said, "Those three guys over there said you'll pay for this." And they were three people off the local rig...

Evasiw describes the hotel fire:

...As the hotel was burning ... they were trying to save all that beer. So, when the first truck - a half-ton - backed in there, and they loaded up (ha ha). A local fella looked at that - and he didn't say anything - he just backed his truck in there ... (ha ha ha) Did they have a beer party after!

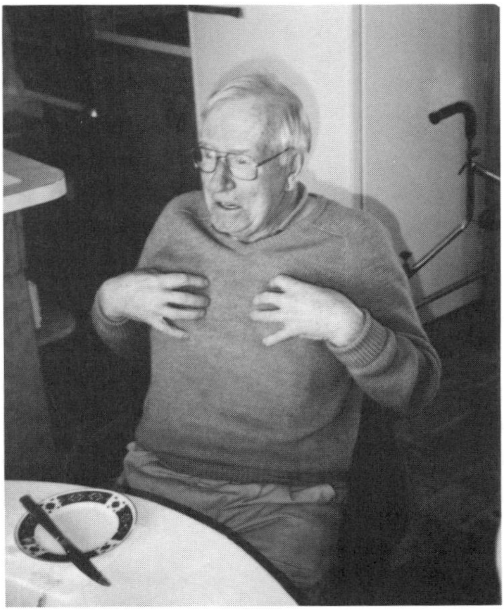

Denis pontificating in Anne Key's home (Redwater).
Furnished by Author

Mary Sawka, widowed with four young children, had to work - first, at The Greasy Spoon during the boom. Then, when the new hotel was built, "...Kitchen ... I did cooking ... I worked 'til ... I quit when I was 65." What devotion: 20 years slaving over a hot stove!

Maxine Kerber has seen a lot of action. Currently manager of the Redwater Hotel (built in 1949 to replace the one that burned down in 1948), she had worked there 20 years ago in a very junior capacity.

John Hunchak had grown up in the area. His operations were family oriented. One of these was the Egremont hotel, which is still standing. When the Redwater Hotel burned down in 1948, shortly after the discovery, John's hotel took over a lot of the business. He acted as barkeep, along with four other waiters, and had the job of tapping the kegs. He ran taxi-cabs out to the rigs and back and saw to the construction of the Roxy Theatre in Redwater.

Paul Greenwood recalls the morning "kaffeeklatsch".

> ...The headquarters for the Morning Coffee Brigade was the Redwater Cafe run by Pon Lee. You either went there about 7:00 a.m. or about 10:00 a.m. and, invariably, anybody you wanted to see would drift by to play a little game of 'birds in the bush' for the coffee...
>
> The "Commerce" Bank manager was an entrepreneur, but his only problem was he liked to do a little too much of his business from his favourite table in the Redwater Hotel with sometimes disastrous

results with some of the more antagonistic loanees - if that's the right word...

John Chekerda, retired superintendent of Redwater Water Disposal Company, "We had a store in Redwater - hardware, groceries, and meats. And there was hardly any place to eat, so Mum used to cook meals for the workers in the back of the store."

On the temperance side, Jim Lukes explains their move to Redwater to set up a drug store. He thought they'd make a go of it in Redwater. Jim recalls their first trip, hitch-hiking from Edmonton. "We went to see Mr. Phillip Horbal ... He had a store there ... people sitting on Coke boxes eating garlic sausage and drinking coke ...

Jim recalls Lauder Nowers coming into the drug store and assuring him that Imperial Oil employees would pay their bills, implying that Lauder would take it out of their hides if they didn't.

Steve Bleakley recalls:

> ...We stayed in the men's 'batch' as the (Bird) houses weren't finished until just before Xmas (at least ours). After work, as well as spending time in the beer parlour, we made our own meals (there was a hot plate in one of the geology shacks). Only rarely did we eat at the Mess Deck, which at one time was the only cafe in town. There was a lot of poker played in the batch and, as well as playing a lot of crib, Jack Nesbitt and I used to take on Bob Ings (Personnel) and Bill Hosegood (Accounting) in bridge for 10¢ per 100. We'd play 'til all hours of the night. There's a hole in a wall (about fist-sized) in one of the rooms in the batch, apparently put there by Jack one night when he was playing cards. I wasn't there...

Steve tells yet another story: Jack Nesbitt and Bob Elliott had been painting the town red. Jack thought he had been short-changed by about $20. So, from the lean-to behind the Mess Deck, they liberated a frozen turkey and walked it over (it was large!) to Steve and Diz Bleakley's house. By this time, it was 2:00 a.m. Both of them were hungry and, naturally, they wanted breakfast. Diz turned them away, but allowed the turkey to be taken downstairs ("And then get the hell out of here!"). Diz's version recalls it being New Year's Eve, when she hosted a party that went on until 5:00 a.m. She was awakened an hour later by the same two bozos pounding on the door. The chain of events that follow is the same.

By morning the head had drooped over and the wings were starting to thaw, so Steve put the bird in an apple box underneath the stoop. And there it stayed, frozen, until early spring. That was when they decided to eat it. Gunnar and Rose Haugrud, from Edmonton,

L to R., Deacon, Bleakley, Nesbitt and MacRae.
Furnished by S. Bleakley

were among the dinner guests. That was the time when Rose locked herself in the bathroom and couldn't get out. It is not clear whether it was in Steve's house or in Jack's, but in any case, Rose had to be rescued through the bathroom window. The party went on without any further ill effects and the liberated turkey was demolished.

Steve Bleakley mixes geology and gin in another vignette.

> ...I recall one of these 'high-priced-help' people from (New) "Jersey" coming up (Mose Knebel), and we had come in to Edmonton for 'educating' and, I think maybe, a dinner. I know both Ernie Shaw and Doug Layer were there. At any rate, about midnight, Hec MacRae and I and either Jack Nesbitt or Bob Elliott decided to visit Mose in his hotel room. He appeared at the door in his PJ's and immediately shooed us into his ante-room and went back to bed. In this room was a plentiful supply of different types of liquor, to which we helped ourselves. We opened another door and found both Ernie and Doug in bed asleep! Eventually we made our way out, but hadn't gone too far when Hec remembered he had left his hat there - and it was a hard-to-come-by size 7½! He wasn't about to lose it, so he had to go back and get Mose out of bed one more time! As I recall, Shaw had some wry comments to make the next day...

The "wheels" had their own wild parties. Jack Harvie recounts:

> ...In the fall of '49 - we had the new office in place - Imperial directors came up, as they usually do once a year. Scottie Jamieson, Edmonton

Marketing man, was in charge of the arrangements.. I know there was George Stewart and Jack White and Bill Twaits.. Well, that was a disaster because, you see, they had an open bar ... and the directors came in late. By that time, everybody was loaded. George Stewart tried to make a speech. Nobody was listening, but he soldiered on (laughter) ... It was just a complete disaster. The next morning - they'd arranged to come out to Redwater and tour the townsite - so, we get out there ... and the Grey Cup was on. So, Bill Twaits - an old football player, you know, he said, "What the hell. We'll take a quick tour and listen to the Grey Cup game." So that's what we did. And that was the year Montreal beat Calgary, I think 28-14, so we all sat around (laughter)...

Recreation was not confined to beer parlour elbow bending. Imperial Oil was fully cognizant of the need to provide other facilities, notably a curling rink. And Campbell Aird, the draw master (also production superintendent in his spare time), saw to the construction of a rink paid for by Imperial Oil. (When one thinks of how little their crude cost them, they could have built a Saddledome.)

A bonspiel in the early 50s. L to R, Si Cormack, Leigh Constable, Millard van der Grift, and Bud Kelly.

Furnished by Si Cormack

Bonspiels, especially with neighbouring towns, were the centrepiece of rock throwing. But these events had their hazards, iron peddlers furnishing bottles of whiskey at each end so that none of the curlers would go thirsty. Bleakley gives his impressions:

> ...When they built the curling rink - I think it was the second winter I was there -anybody who had even SEEN a rock before was an automatic skip! The calibre of curling wasn't all that great, but there were certainly some memorable bonspiels - often using ice at Egremont and Bon Accord and curling all night. At the first bonspiel (and quite a few afterward, I guess), the ladies of the town served a TERRIFIC banquet with turkey and all kinds of Ukrainian food. And the service companies had their reps there with all kinds of liquor under the table. I remember, after one meal, we (Nesbitt, Jack Turner, Jim Henderson, and I) had to play the station agent and his rink. Everybody had over-imbibed so much (except Jim) that the score after about six ends was 1-1 as I recall. This was not because of good curling, but because nobody could get a rock into the house. We even changed skips with no better results.. I think Jim went home after about three ends - in disgust! Many of us took in the spiel at Redwater for years after moving to Edmonton, and even Calgary...

The other focus of social activity was the Legion Hall, where the vets would gather. Jim Evasiw was a strong supporter - as was Victor Kerber, whose widow describes him,

> ...My husband was a hardworking man, and, being a Legionnaire, he and his roughnecks practically built the Legion Hall, here, from the ground up ... If you wanted to get any place - in this town, especially - you had to go that extra mile. My kids all did well in school ... all very athletic...

Schools had their social functions, as Alex Chaba recalls the respect which the otherwise wild roughnecks displayed when they attended the one-room school functions in the area without being invited.

Women were the mainstay of the community, but so many had been nomadic - moving from one town to another - that they had developed few bonds. And young mothers were out on their own, some for the first time, with few social graces, timid yet wanting to be accepted. Ruby Nesbitt, Jack's first wife, was a shy, self-effacing bride, the exact opposite of her husband, rambunctious, always on the prowl. She found it difficult to fit into the community but was fortunate to have made fast friends with her neighbours, the Bleakleys. It was into this atmosphere that Dean Hunter injected her unique brand of cheer - notably, get-acquainted parties - "come as you are" as Anne Key recalls.

Kay McCaskill:

> ...It was Dean Hunter's idea to form an oil wives club, as families were being moved around so much in those days and the wives - usually with young children to care for -were often very lonely. So Dean assembled a group of six or seven of us ... The next year we were moved to Calgary, and I organized the Calgary chapter ... Dean was a very special lady...

Alice Lukes has a precious memento in the form a good-bye card which was signed by all of the Oil Wives, as well as a full-sized photo of the party at which all the gals arrived "come as you are". It clearly shows the affection that they held for Alice. She reminisces,

> ...Well, you know, it was the beginning of Oil Wives ... It seems like such a long time ago, but it was a very friendly group. And because I have the picture, I guess, I remember that evening of us all being together. And I realize now that the reason why we had that 'hard time' party is a lot of us girls didn't have clothes for a special party. A lot of girls wouldn't have come, probably. But to come as 'hard time' made it a fun thing. The ladies who organized that were very smart ... they made us so welcome and created a community atmosphere that probably wouldn't have existed, even though people coming into the drugstore and those you met on the street were very friendly. And the fact that Imperial Oil helped so much - with the curling rink ... and to put in the water for the town was really special...

Bleakley, geologist, adds his own perspective:

> That winter was bitterly cold. We'd leave the vehicles running for hours on end - both to keep us warm and to make sure they were mobile. The cost per mile got pretty high at times! When the wives moved out (to Redwater), I remember I had to hook up our gas stove ... as we had company coming for Christmas dinner! Poker and bridge games at the batch got much fewer and farther between after the wives moved out...

But Maxine Kerber, manager of the Redwater Hotel today, sheds a different light on the community. Her husband had worked for Beta Well Servicing.

> ...I hate to say it, but I found that in most places, people that already lived there were real snobs. They thought the oil people were really second class citizens. Even here in Redwater, if you worked for a big company like Imperial Oil, you were okay. But if you worked for a well servicing outfit or a drilling rig, boy, you came from the wrong

> side of the tracks! There was a very black and white distinction. In school, the kids felt it, even...
>
> We didn't work for Imperial Oil, so we didn't join (Oil Wives). We were the roughnecks' families, you see ... We've put on some fabulous entertainment for people, at the hotel, and the town of Redwater just doesn't show up. I think people feel the hotel has the name of being a place where drunks hang out.. You can look around and see, yourself, it's not true. We have a nice clientele.. a lot of them sit and drink coffee.. People don't get drunk anymore ... the way they used to. When I worked here 20 years ago, people got drunk!

World War II created a temporary need for female employees, but it was generally frowned upon, says Kay Jeffery.

> ...They didn't want you to work. When the fellows came back from overseas, it was frowned on. They didn't want the girls to stay on after they were married. I worked for a year and a quarter in Calgary Office, and they had said that they would have to let me out because of the fellows coming back - but they never did. But, when I was transferred to Redwater, it was understood - if you had a working husband, you did not work. You were supposed to be home and looking after the house and family...
>
> When I was in the Calgary office, Peggy Donnelly was on the switchboard. I didn't know Peggy that well, but there was a cute story about it: There was a VIP up from Toronto, in the main office, and he phoned down.. wanted an outside line.. So, Peggy said, "Fine". In the meantime, he's talking to the people in Walker Taylor's office.. and all of a sudden, he said "Jesus Christ". And right in his ear was, "No, it's Peggy." That broke everybody up for a few minutes...
>
> I can remember ... 'Black Mike' Haider coming out ... He was a perfectionist ... When he gave dictation, he wanted every word exactly as he dictated it. I don't think he worked out of Calgary, and I don't know why they brought him out there. Dictation wasn't my long shot ... I had to take Pitman dictation from him once - I think it was the most terrifying day of my life! Oh, he was very nice, but he was so exact that you didn't change an 'and' or a 'the' in anything ... I think I only had to rewrite it once...

Land values are the surest barometer of economic activity, and Redwater was no exception. Alex Chaba remembers land prices before the oil boom. No one wanted the unarable land east and southeast of Redwater which could have been purchased for as little as $400 a quarter (by paying tax arrears). Much of the land was muskeg, pine, sand and pasture. Blueberries were the cash crop: $0.25 a pound.

When oil was struck on the jackpine-muskeg acreage owned by some farmers, they immediately sold their lands (no freehold oil rights in Redwater) at a windfall price of $16,000 per quarter and moved on.

The story of the very first attempt by Wilbur Anderson to buy land has already been told in Chapter 2. When the farmers realized there was another cash crop on their land not subject to frost, drought or hail, some of them bought all kinds of acreage to ensure they would get surface rental for well sites and batteries.

Kitty Wilkins, a native of Redwater, has made it her business to keep on top of surface land dealings with the companies. And why not? She has had to renegotiate with Imperial Oil on her quarter section several times, and she is still receiving rentals.

> ...We had a salt water spill ... went into the creek and killed trees ... You get a bad landman, he won't pay for damages ... We just took their word for what they said.[5]

Kitty knows about outright sales of farmers. One of her neighbours cried when he realized what he had done (love for the soil conquers all). Nevertheless, the farms that Imperial and others purchased created weed-cutting and fencing[6] problems. The farmers had to be paid to cut what used to be their own weeds! Fences around well-sites were routinely knocked down. The father of the foreman of Imperial's Transportation Department got the job of mending them!

Mrs. Evasiw describes how her father, Mr. Zeniuk, acquired land.

> ...They moved to Eastgate, just about five, six miles from there. My Dad bought this land in Redwater, where the old town site was, just about a hundred feet or so. So when they discovered oil, they (Imperial Oil/ Devon Estates) wanted my Dad's adjoining 160 acres plus this 130 acres, here, where the town was...

Jim Lukes remembers his encounter with Mr. Horbal when he was looking for a place to set up his drug store.

> ...He didn't really want to sell. We had sent out letters to some of the property owners, indicating we wanted to start a business. So when we went there, Mr. Horbal pulls out this letter, "Here, I got a lot of people want to buy my property ... here's a letter from R. C. Brown in Calgary." And we said, "Well, that's us." The reason we were able to

---

5 Jack Nesbitt, one of Imperial Oil's geologists, considered their treatment of the farmers as being equitable, "It was mostly jackpine and slough, so in a sense it was something like revenue in lieu of crops."

6 Conservation Board regulations required attention to these.

> buy it is that his wife had a big garden plot right on main street.. and she didn't want to sell -and she was a lot younger than him - and I think he wanted to show her who was boss...

Ray Haffner, B-A's foreman (transferred from Stellarene Refinery at Coutts), organized the company community. Each of the occupants bought parcels from Roshko - eight lots on the 3⅓-acre tract (Ted Court thinks he paid $150).

Ed Lakusta confirmed his father was a tough s.o.b. Mike refused to sign any contracts that had no minimum acreage clause.

> ...He was tough, but ... he recognized he was signing a lease to probably last beyond his livelihood ... that if he signed it and didn't cross all the t's and dot all the i's that somebody could come along and change the deal. At that time they were paying $100 per acre for subsequent years..My father was concerned later they'd reduce the acreage to maybe one. He didn't want them on his property for $100, but $400 seemed fair. I think the other thing, he recognized ... you know, the agricultural and farming machinery was relatively small at the time - pulling 10-ft disks ... and cultivators. What he saw coming ... was 30, 40-ft disks, cultivators, big machinery ... He really saw a well in every 40 acres as a nuisance ... for efficient farming operations ... wondering how he was going to be compensated if he signed the leases today...

Helen Lakusta was asked about her husband being a tough person in his dealings with the oil companies, and the answer was straightforward.

> ...Yes, he was always trying to make as much as he could ... nothing wrong with that ... They're on your property ... they're right behind your house ... they have to pay for it. But they weren't paying too much at the beginning...

An outside source claims that, on one occasion, Mike was served with a court order when he refused entry to his land.

Bill Allen relates another encounter with Pete Yarmula. Bill had been given the job of acquiring easements and rights of way from the river north to Redwater in the course of preparing to lay the water line to the community. He had progressed rather well until he got to Yarmula's property just east of town. There was no way Yarmula was going to sign anything. But Bill knew Pete's weakness, so off they went to the beer parlour. After an appropriate number of beers, Yarmula was in no condition to resist Bill and signed the release.

Gordon Hawkins tells how he and Hubert Somerville were called out to examine the complaint of a farmer who was blaming his house damage on shot-hole drilling, shooting, and the resultant vibrations.

When they approached the house, they noted nothing out of the ordinary. It was pebbled with glass in the stucco. They then went downstairs into the dugout basement, where the farmer showed them cracks in the concrete which were supposed to have been newly created. On closer examination, Gordon noted the number of spider webs that were covering these apertures. He can only conclude that these were old cracks or else the spiders had been working overtime. When asked where all the glass that had fallen off the house was, the farmer could only say, "My turkeys ... my turkeys have gobbled them up" with severe consequences.

Despite these outrageous allegations, the oil company involved felt it politic to compensate the farmer -another example of the oil industry falling over backwards to keep everyone happy.

Ernie Fruman, surface landman for Imperial Oil, fled Yugoslavia in 1945, and finally worked his way to Medicine Hat in 1948. W. F. (Bill) Clark, Ernie's supervisor, confirms this and adds that he had command of several other European languages, including Ukrainian. This aptitude, plus his legal training in Europe, gave him an advantage in his negotiations with farmers. Clark remembers his clear scroll-type writing. "You could always depend on him." When he went out to negotiate he was completely reliable.

The most interesting story was his meeting with a Mrs. Hally. The story is told by Nancy Chaba. Mrs. Hally, a widow, lived in the Amelia area. She had a problem: a beaver lodge which she said had not been there "before the oil companies came in." She was persuaded to buy a bottle of booze to ply the two landmen, Smith and Fruman, and get them to give her extra compensation for the "Imperial" beavers which were flooding her land. But Smith and Fruman knew that the lodge had been there for a long time. Mrs. Hally had always to have an interpreter and she got Nancy's dad, Andrew Yachimac (no relation to Anne) to intercede. The meeting ended with an empty bottle but little, if any, extra compensation. She continued to hound Imperial but to no avail despite complaints to her neighbours.

Walter Hrynchuk comments:

> ...It made a lot of people richer by having leases drawn up ... setting them up for years. They paid a decent fee, and then there was a yearly payment, so it kept the farmers going well. And they were able to live better - some had bought new vehicles, and stuff like that. So, it made everybody happy - the wells coming in and stuff - because they got payments and stuff like this. And another thing is that lots and lots of

> people had jobs. The town prospered well - stores, restaurants came up...
>
> And there was people phoned Imperial Oil.. and the farmers who were against having oil wells drilled here because they said it was going to kill the trees ... the grain won't grow ... nothing will grow - gardens or anything ... everything will rust away - and stuff like that. So people that came out from Imperial Oil explained that nothing is going to happen. And we see now, nothing did change. And the good thing about it was, I would say, about 75% of the people that were living here were working in the industry that Imperial Oil produced here...

With the delineation of the field by 1952, crown reserve sales ended and drilling ceased. There were still many hands looking after production but, as automation advanced, their numbers shrank until today large, remote areas are supervised by pushing a button or giving a voice command.

Anne Key muses, "When the field was drilled out and the rigs had left with the transients, the community matured ... and the people who stayed were of a different calibre."

John Hawrelko, commenting during a 1992 interview:

> ...Well, I look at the National Energy Program with some equanimity. I'm not exactly hung up on this idea of losing $65-$70 billion. It may be a fact. It may well be just a statistical distortion. I note that the Conservatives who got into power took a couple of years before they finally got rid of the N.E.P. It was a good milking cow for the politicians in Ottawa. From my point of view, as I look at the story of Redwater, the oil was in many ways a god-send, because the land here was largely marginal. Much of the land west and southwest of here is very poor, mostly sandy soil. So, when the oil came in here, suddenly the roads improved. And Esso comes out as a good corporate citizen over the things that they did in the area. They may well have bombed out on the land that they acquired, but they seemed to have had a good reputation in the town. They were looked up to by a lot of people. They appeared to be fairer with the farmers than were the other oil companies. I really haven't much to say against these oil companies. If it hadn't been for (them), we'd still be riding over difficult roads, I'm sure. And this town would likely have disappeared because there was nothing to sustain it. The future? I hope industry will come in to this area. I don't think this town will die just because the oil field is on its last legs. There is, I think, hope here for, say, OSLO[7] and maybe

7 Other Seven Lease Operators.

some subsidiary industries, if the province can disperse industry ... and allow small towns to continue to be viable...

Ed Lakusta, in summarizing,

...I look at Redwater as an example of how you can have coexistence between an industry ... and an agricultural community ... They both support each other ... When I look back ... over the last 40 years, I think it's been great for that whole area...

The Redwater oil field has been good for the total community within the field and a large peripheral community around it ... and the province in terms of royalties, and the municipality in terms of taxes generated...

PROFILES

... for posterity's sake ...

JULIUS ABZINGER

Julius Abzinger is Hungarian, born in Czechoslovakia and a graduate of a Hungarian institute: Jozsef Nador University in 1936. His claim to fame is that he knew George Bannantine in Rumania. When the Communists came into power Bannantine was thrown into jail for one week, but the powers that be managed to gain his release. Julius weaves a chaotic account of how Bannantine came to Canada and, in an equally confused way, how he himself actually came. He claims he was wined and dined by Esso International somewhere in Italy, then spent about 20 months in Switzerland. After crossing the Atlantic in Steerage, while his wife luxuriated in First Class, he took the train from Halifax to Toronto, ending up in Mike Haider's office at Imperial's headquarters: 56 Church Street. (He described Mike as an Austrian with a very tough temperament.)

Julius then travelled by rail to Imperial's Calgary office, 300 - 9 Avenue S.W. and met R. B. Curran (whom he describes as arrogant). He recalls receiving his orders from a Mr. Martin in Personnel. (Did he mean Bummer Stirling?) From there he was sent to Edmonton, where he was warmly reunited with George Bannantine, who hugged him. His journey ended in Redwater, where he lived in the barracks and commuted back and forth to Edmonton on weekends to see his wife.

Julius said that his first job was cost control; then he switched over to Drilling Engineer, working under Eric Connor. He drilled 400 wells in Redwater, at Excelsior, Bon Accord and even Golden Spike before moving to Chatham, Ontario 2½ years later.

His account is probably more colourful than substantive, but there is no question he was at Redwater, even if his exploits are somewhat overstated.

IVONE ST. G. BURN

Ivone St. G. Burn had been out of the industry picture for a number of years due to failing health. However, even after his death, in 1992, his high principles live on in the memories of his compatriots of 40 years ago. His son, Ian, provides a most admirable memorial.

"Ivone St. George Burn was born August 26, 1906 in Lundbreck, Alberta, and was raised on the family ranch. The neighbouring town of Burmis was named after his grandfather, R. H. Burn, and a fellow rancher, J. Kemmis. Ivone moved to Vernon, B.C. to attend high school and the University of British Columbia.

In 1929, he joined Imperial Oil Ltd. and worked on geological surveys with Dr. Theodore (Ted) Link who, over the years, became a close family friend. From his early surveying, he worked his way up to the head of Imperial's land department. His early background with rural life gave him an advantage in being able to understand the farmer's and rancher's perspective in land agreements. Ivone was one of the many important players in Leduc and Redwater.

Ivone Burn, then-member of ASPG, 1946
ASPG Roster, 1946

In 1948, he left Imperial to manage the land department for Pacific Petroleums Ltd. and two years later accepted a similar position with Canadian Delhi Ltd. He next joined Scurry Oils Ltd. as vice president and manager of the land department until he started his own firm, Burn Oil Land Services Ltd., in 1953. Through this firm Ivone handled many land negotiations for a number of the smaller oil, gas, and pipeline companies.

After retiring in 1972, Ivone and wife Phyllis moved to Victoria, where they lived until his death on October 26, 1992. He is survived by his wife, son Ian in Edmonton, and daughter Betty Jo (Sudul) in Sidney, B.C.

A quiet, capable and conscientious man, Ivone Burn played an important role in Alberta's oil industry. He will be missed by his family and all who had the privilege of knowing him."

THE CHABAS

Alex Chaba is an antiques' collector, but if the truth were known - and the locals agree - he is a packrat par excellence. The bursting out-building on the Court property is mute testimony to his acquisitiveness. His wife, Nancy, has been swept along with his determination and accompanies him to Edmonton's Westview Inn, every Sunday, to meet their counterparts and to swap and upgrade their collections. No ordinary acquisitor, Alex and his associates were successful in sidestepping authority (he's good at that) to have a locomotive moved to a Stony Plain museum site which he owned and operated from 1960 to 1967. Fort Edmonton subsequently purchased it.

Alex and Nancy Chaba with some of their treasures.

Furnished by Author

Alex was one of a family of six born in the Eastgate district southwest of Redwater. His parents, Fred and Jo-Anna (nee Pysmany) homesteaded there in 1912. Alex recalls his early years, spending his time at Eastgate and Ufford schools. He attended the University of Alberta, took Education, then went teaching school.

Chaba reluctantly left the teaching profession, which paid $98 a month, to work at the oil business for Imperial Pipeline at the new loading racks west of town (on a Pasemco parcel) for a wage that was triple his salary. Other episodes included being with Petrofina in many capacities. Alex worked for Majestic Pipelines laying salt water disposal gathering lines and replacing those that were starting to leak. However, he found time to enjoy life going out with some of the visiting "wheels" to hunt ducks and upland birds. He was already showing a propensity for rubbing shoulders with the people that mattered.

Nancy (nee Yakimec) and Alex were married in 1953. The wedding must have been the social event of the year - her dad, Andrew Yakimec, invited everyone to the two-day wedding. Andrew owned seven quarters, at least one of which had been drilled by McIntyre & Webster. Her parents sponsored Nick Rozak.[1]

Behind all those old license plates and kitchen scales rests a crafty mind - one that's always looking for the best deal. Maybe that's why Alex never became a farmer like his brother, Walter.

Walter Chaba has a packrat nature too, but not so sophisticated as that of brother Alex. Walter is a friendly son of the soil, huge coarse hands roughened by much toil. One can detect that his roots are in the Ukraine; in 1992 he'd made a pilgrimage to Ternopil, the birthplace of his father. While there, he relived the stories his parents had told him of oppression by the Polish lords on their tiny tract of land.

Walter's home shows all the signs of his pack-rat nature. His doorstep is cluttered with masses of bolts and nuts which he proudly bought at an Edmonton auction sale. Walter welcomed the author into the house, the ubiquitous kitchen table being the heart of the cluttered menage. Over coffee, Walter gave a brief history.

Walter's education at the Eastgate School ended at Grade IX, and he helped his father, Fred, on the farm. He remarked, with a faint puzzled smile on his face, that he could not remember his birth date, explaining that he "had four by-passes and was dead for a minute and a half."

In 1945 Walter married Theresa Scragg, and they have two sons and a daughter from that union. Theresa, who had just returned from town, took over the conversation at that point: "You needed a job to keep farming. You couldn't cover expenses without a job." `Walter went to work for Imperial Pipeline in 1950 at 87¢ an hour; this was

1 Also profiled in the Blue Pages.

increased to $1.25. His jobs were varied: bull gang, gauging tanks, loading tank cars at the Redwater siding, even line walking and some work in a carpenter's shop. He retired from Imperial in 1963 and went back to farming full time.

After the children grew up Theresa returned to the classroom but has since retired and leads a simple life with Walter. Here was a couple coping, not really sure of what lay in store, with those unfinished/abandoned projects in the yard. The highlight of this particular day was the arrival, in the mail, of their Old Age Pension - one of the few outside sources they could depend on.

THE CHEKERDAS

John Chekerda, recently retired field superintendent for Redwater Water Disposal Company, tells his life story: Born in 1931 and raised in Redwater, his family had come from the Ukraine about 1904 and settled on a farm. John remembers as a young person, "We were going by where the derrick was drilling (Imperial Redwater No. 1 discovery) ... The derrick was all brown ... I wondered what the heck happened - they couldn't have painted it overnight. Then you found out they had hit oil." He ties the Redwater discovery with his Grade X year.

John worked for a while in the bunkhouses that his family rented to Sparling & Davis, later joining Northern Oilfield Ditchers, contractors for the Redwater Water Disposal Co. He "got on steady" with the latter in 1951. "They had three wells when I started."

Now taking life easy, Chekerda can look back with satisfaction, having worked his way up from a common labourer to supervising 50 wells and a small army of men servicing the operation.

Larry Chekerda, John's son, is a bright, young, aggressive technician at Imperial's gas plant in Redwater.

Larry first saw the light of day in 1956 on a farm near the town, eight years after the discovery. He attended Radway elementary school, then continued to Grade XII in Redwater. He proceeded to an instrumentation course at NAIT, graduating in 1976. After a few months with Home Oil at Carstairs, he took a job as instrument technician at Imperial's plant. He was later transferred into Special Projects, looking at measurement and automation and also investigating improved well testing.

It was a sunny September day in 1991 when the author interviewed Larry beside his pride and joy. The abandoned, rusting derrick had been removed from Imperial's yard, refurbished, and erected on a site in the town park with Larry as the driving force, who wouldn't take "no" for an answer.[2] He could grasp the historical significance of this derrick because it was the one that had drilled Redwater No.1 on the Cook farm and was moved up to drill No.4 at Egremont.

Says Larry:

> ...There was a large 16-inch pipe through the middle (of the derrick) at the plant and the derrick was a support for the pipe, which was the flare stack ... The derrick was no longer needed, and Esso put up a little free-standing stack ... Redwater Town Council was approached and they said, no ... no interest in putting the derrick up in town. I couldn't believe it! I got a few people together and we went to a Town Council meeting, and I said, "What's going on here? How can you let something like that go?" Council said, "Well, if you've got somebody willing to take the project on - fine, we'll keep it..." That's when I got stuck with it.. and the challenge of getting the derrick re-erected...

The deadline date was February 1989, and the first concern was to raise enough money; otherwise, the derrick would go to Devon. But by the summer of '89 the derrick project was completed. Larry, with his penchant for detail, made sure that the girts were sand-blasted and painted before the derrick was moved.

> ...We got the whole thing done for under $30,000 - painting, blasting, welding, cement, and foundation were contributions ... Poole Construction put it up for us. We had some of the original workers stop by...
>
> Associated Engineering donated their time to fabricate the massive concrete corner posts which support the derrick ... It may appear to have been over-designed, but it is a good feeling to be on the safe side and not have to worry about it falling over...

---

2 See Appendix I for list of donors.

CECIL EDMUND CHESHER

Cecil Edmund Chesher, geophysicist-cum-entrepreneur and a native of Petrolia, Ontario, had graduated from University of Toronto in 1939 with a degree in mining geology. He had returned to his home town, unemployed, when he overheard members of a Heiland seismic crew in the local restaurant. They had come up from Louisiana to shoot the area. He talked himself into a job and returned with the crew.

Later that year he married Margaret Allan (alumna of Ontario Ladies College, Whitby, whom he had known at U of T) in San Antonio, Texas. He is said to have worked for Heiland, on contract with H. L. Hunt in the U.S., at zero salary. When he had proved his worth, he was put on the payroll. Wilf Baillie recalls Cec's pledge to be a millionaire before age 30, and he was! His obituary in The Albertan of October 7, 1975, credits him as "the discoverer of two major oil fields" (Jumping Pound and Redwater).

Chesher is believed to have come to Calgary in 1942 as a Heiland employee. He was living in the Palliser when his first child, Margaret, was born in Toronto that year. Gordon Hess, who went to work for Shell in 1943, says Cec's crew was the first one to winter in Canada - the ease of traversing frozen ground being the main attraction.

There are no records to prove or otherwise, but Gordon says Chesher must have ramrodded the Jumping Pound area conducting both refraction and reflection plus side shots. Don Redman, then a Cal Standard junior engineer, recalls Chesher running a velocity survey, April 1943, at Ross Lake dry hole.

Ian, his son, says his father was sent by Shell to Cal Tech and Colorado School of Mines for upgrading courses. The ultimate result is believed to have been the discovery of Jumping Pound in December 1944. Francis Reeve, founder of Commonwealth Drilling and Dominion Oilfield Supply, is said to have bankrolled Chesher, encouraged by Reeve's wife, who had taken a shine to our hero. When he started to cash in, Reeve was upset at Chesher's driving around in a Packard straight-eight status symbol ("too young").

Cec incorporated his own firm, Canadian Exploration Company on August 31, 1945. On February 4, 1946, he formed Chesher Exploration Company. In 1947, Cec was listed as living at 1718 - 10A St SW, Calgary. D. P. McLaws served as director of Chesher's company. Records show that the offices moved to the Fisk Bldg, 342 - 7 Ave SW, where the Bank of Montreal skyscraper now stands. Steffie Negrich recalls working for the company there.

Before Leduc, according to Gordon, Cec owed so much at the Palliser that they couldn't kick him out, especially since he had his family with him. But Leduc changed all that. Some time before 1947 spring break-up, Chesher shot Leduc and confirmed what Ray Walters already knew. Cec was now ready to contact some real clients.

As soon as Redwater hit in the fall of 1948, Cec moved his crew out with Gordon Hess as party chief. They kept busy until all the crown parcels had been disposed of in 1952. Cec concentrated on shooting the edge tracts, knowing the central leases were beyond his clients' means. Details on the Crown Reserve purchasers and rewards are to be found in Chapter 6 and Appendix XVIII.

Chesher also drafted up a contour seismic map of Redwater and started to flog copies at $10,000 a pop. According to Bill McLaughlin, Pacific's geophysicist, when Frank McMahon heard that the maps were being peddled, he immediately called Chesher in and paid him enough (we'll never know how much) to take the map off the market.

Cec and his first wife, Margaret (deceased Jan 28, 1949), basking in the Banff sun.

Furnished by Ian Chesher

In return for this "exclusive," Frank granted Cec a 1% gross override on all the crown reserve parcels that Pacific Pete and partners would buy.

Margaret died in childbirth, leaving three small children. Cec married Joyce Willins, by whom he had three more children. They subsequently separated, and Chesher married divorcee Madge McKay. Margaret Allan, Cec's sister-in-law, was less charitable, "There were many other 'acquaintances'."

Chesher soon started to live the "good life." He became a member of the better clubs, raising and breeding thoroughbreds - likely in partnership with Max Bell and Frank McMahon. He was president of the Stampeders Football Club in 1952. He was one of the original promoters of Big Inch Pipeline Company, "The first producer of large, 36-inch pipe for gas and oil pipelines in western Canada."

Chesher's associates state that he was totally out of geophysics when he died in Palm Desert, California, in October 1975 of a heart attack at age 57.

THE COOK DYNASTY

Hilton Cook is synonymous with Redwater oil, because it was on his SE/32-T57-R21 W4M section that Imperial Oil drilled their discovery wildcat in the fall of 1948. Company PR and national magazine articles publicized Cook as a model farmer, causing no trouble over surface rights. Hilton mistakenly claimed that he missed obtaining his mineral rights by one year (1905). He was actually 19 years late, 1887 having been the cut-off for crown homesteaders. John Shaw, Jr., Hilton's son-in-law, would later negotiate higher surface rentals on behalf of the family

Hilton and Christina's fiftieth anniversary, 1953. L to R: Back Row - Maurice, Jack, Bud, Percy, Jud. Front Row - Margaret (Parsons), Winnifred (Shaw), Rita (Astle), Alice (Holst), Effie (Berglund).

Furnished by R. Astle.

Hilton's father, Samuel, was one of the first - if not the first - "English" settler to build his house at (what became) Cookville near the North Saskatchewan River. Samuel's first wife, Amelia, is memorialized by having a school and one of Imperial's lease blocks named after her. "Memories," a 1972 Redwater history book, provides

Cook House (Riverside). Rita claims the ghost of her grandfather is looking out through the window. A small, child-like effigy cannot be explained.

Furnished by R. Astle

Samuel and Amelia Cook.

Furnished by L. Key

their portraits but little else: Samuel - stern, affluent looking; Amelia - beautifully groomed, with a haunting, sad mien.

Most of the author's information comes from Rita Astle, Hilton's youngest daughter. She describes her grandfather's house: "... Where Grampa used to ferry the settlers across ... I visited the site ... entered the house and went up the stairs ... bits of wallpaper clung to the old walls, but there was no furniture ... I sat on the floor and tried to think about how wonderful it would have been if I could have only known them, even for a little while."

Rita grew up on the farm and attended Pine House School (2½ miles distant). There was relative affluence in the twenties, measured by the first gramophone and, according to her, the first Model A Ford. On top of this, her father took great pride in "a big, old Marconi radio ... it ran on a car battery ... The rest of us were not allowed to touch it ... His favourite program was William Aberhart's Sunday morning sermon." Rita also remembers her father and a Mr. Roberts (Scott?) bringing two loads of children to listen to King George V's silver anniversary speech in 1935. Large numbers of relatives would descend upon the house and her mother would try to figure out how many mouths to feed. "...They never went home hungry."

"Once the Depression hit, things changed considerably ... no more car ... My father was too proud and stubborn to go on relief, but somehow we managed to survive."

Rumours had started about possible drilling for oil, but Hilton did not become involved until the day Bill Smith, Imperial Oil landman, visited the farm June 9, 1948, for the purpose of signing a 25-year surface lease agreement. This included the road and the wellsite (24 acres) at $65/acre for a total of $1600.

Jud Cook, one of the sons, remembers his father's great interest in the driling activities. "The old man used to go down that little bit of a hill and stand beside the derrick day after day ... It was the following morning before he knew they'd hit oil ... Mother went out around seven and said, 'Dad, that derrick is all afire!' The old man went out to see for himself. 'Mother, that's not fire. That's oil. That's black gold." Another son remembers noticing the rig on his father's land was black with oil. "Everyone was pretty secretive. The landscape was crawling with oil scouts ... On weekends there'd be a thousand people out there standing around and picnicking."

Jud said he was the first of the family to work for Imperial. "We were thrashing, and I was pitching bundles ... Fin Lineham came out to the field and took the fork away from me, and offered me $1.20 an hour ... I had no idea what the oil business was all about."

One day Hilton's son, Maurice, brought Gordon Astle to the house for Christmas because, being a roughneck on shift work, he was unable to go home. Rita describes their courtship. "After our Christmas supper, we went in to the living room for a sing-along, as my favourite hobby was playing the guitar. Gordon was enjoying it, but I started to notice that he was dozing off. I rather hoped that he was just tired, and that my singing didn't do it. We were married the following year." Rita was involved with the primitive switchboard in Boston's Store in Redwater.

Gordon and Rita moved several times throughout the prairies during his Imperial Oil career, ending up in Saskatoon for an early retirement.

THE COURTS

Still jubilant over her most recent triumph: Senior Rodeo Queen, this bundle of energy belies her senior years, over 60 of which she has spent in wedded bliss. Pearl Court's ten-plus-one children, numerous "grands" and 14 "greats" round out her extensive matriarchy.

The two love birds (the Courts).

Furnished by Author

In contrast, 80-year-old Ted is low key, but still exudes the romance of a one-time cowpuncher. Their hard, twangy accents tell you they come from border country, both having U.S.-born parents. Until you question him, you would not think he'd had two close brushes with death. The October 1991 copy of The Legion describes

Cpl. Court's close call as a tank driver, 14th Armored Regt., Calgary Tanks in October 1943. "We hadn't gone more than 100 yards into Motta (east coast of Italy) when one helluva bang lifted the tank and damaged the track, which later broke ... attacks by beehive bombs blew a hole in the gas tank, followed immediately by an explosion and fire. All escaped, thanks mainly to the fact the gas tanks were nearly empty." Cpl. Court was to serve, in all, five years - ending up in the Netherlands.

Subsequent encounters were medical: a blood clot in his heart and an embolism saw Ted fighting for his life. As Pearl put it, "We lived a lot of life."

The setting for the author's visit was a huge house trailer in the woods southeast of Redwater along the River Range Road (No. 213). The land is part of SW/22-57-21 and Alex Chaba is still receiving rental revenues from wells thereon. The yard contains outbuildings[3] of every shape and size, several horses, a couple of dogs, and cats, cats, cats everywhere.

Pearl and Ted, still hopelessly in love after all these years, told how they met. Pearl's neighbour's husband had brought this guitar-playing dude in when she was visiting them in Coutts. That was it. In 1936 they eloped from Coutts on horseback - across the Milk River Ridge in January - to be married in Raymond. No money, no job.

Upon returning to "civvy street," Ted was supposedly hoodwinked out of his V.L.A. allowance. But that didn't bother Pearl, because Ted got a job pumping crude oil across the border from the Cutbank and Cat Creek fields in Montana in a five-inch line to load into tank cars for British-American's Calgary refinery.

Despite the 1947 Leduc boom, B-A were singularly unsuccessful in their efforts to find crude. When Redwater broke, by now desperate for feedstock, they bought oil in the ground at the Crown Reserve Sales starting in 1948. Outside of plant employees in Turner Valley, B-A had no production hands. Shortly after the purchases, Ted got his marching orders to get up to Redwater to help with production and shipping of oil.

With the cessation of the pipeline gathering system west of Redwater, Ted retired in 1972 and they moved out to the quarter section they now occupy. Pearl's philosophy is simple, "Our family took our money, but that's what we were for." Pearl is still giving. The author's interview was interrupted by a huge cut (gratefully

3 Alex Chaba uses some of them to store his treasures.

received on a slab of bread) from the beef roast she was preparing for her guests.

Pearl is proud of her prowess as a poet. She compressed her life story into a series of verses which won first prize at a Prince George fair. Another of her accomplishments was as a horsewoman ("best dressed rider and horse" at the 1992 Radway parade). Pearl and Ted helped organize the Gymkhana Club and they continue to enjoy riding their saddle horses. She described the steed she rode in the parade, "...A bit spinney ... no one else can handle him."

THE EVASIWS: POSTMASTERS

Blame it on the Dominion Immigration officials. Those newcomers from the Ukraine, with their (to the officials) unpronounceable names, were too much for the frustrated bureaucracy. The result: an anglicization in spelling with variations depending on the whims of the official. Such is the case of the Evasiws, whose relatives just down the street, spell their surname Ewasiw.

Jim's father, George, came into the country in 1906.

> ...They had to prove up so many acres before they received title, and at that time he could have picked up the mineral rights for another $10.00...(SAK: Homesteaders on Crown land were unable to acquire mineral rights after 1887.)

Jim Evasiw making a point to his wife, Anne, while Jean McLaws looks on.
Furnished by Author

Anne's father, Nick Zeniuk, had arrived in Canada early in the twentieth century.

Jim had served in the armed forces in WWII and was looking for work. Mrs. (Bessie) Walker had been Postmistress and wanted out, so Jim applied and got the appointment. When Redwater crude was discovered the situation soon became impossible.

Jim describes some of the realities of life:

...There was the gambling, as usual. a shack behind Marshall Wells.. They used to call him Peg Leg ... And it was going year-round ... He used to take the rake-off. And he slept in the day time...

The Evasiws are now retired, and Anne comments

...Oh, we're not bored. We're so busy! We do volunteer work ... Tuesday we had a Nifty-Fifty flea market to raise funds for carpet bowling. And on Monday night we were working at the hall pricing all the things. Jim is very active with the Legion - well, I am too - but he's a treasure ... And they have a group, you know. They go and have their coffee and they go and clean their hall. They do their own janitor work, which saves a lot of money. And they're all retired...

Jim adds,

...Like, you see a lot of fellows walking around bored ... retired, you know? I would like to spend one day being bored. I would enjoy it ... Right now, the church -they're having the Bishop coming in ... cleaning the chandeliers ... take down the scaffolding ... I think the youth organizations are doing a terrific job - the Legion sponsors 4H, the girl guides, the boy scouts, the hockey...

One last comment about the post office by Anne. "In the old days, the war vets ran it like a charm. But when Canada Post Corporation took over (and Anne worked for three years under their jurisdiction), the union did what they pleased. There was much less interest in seeing mail go through than in social bantering in the office."

HAWRELKO AND SENYCH

If these two had lived in Athenian times, they would have been titled solons. In the community of today they occupy positions of authority and respect as thought leaders through their demonstrated ability to educate. (What more important vocation is there?) They have been catalysts for change in this incubator of excellence. Instead of insulating their students against the Anglocentric world, they have made their pupils aware of the richness of English literature while at the same time nurturing their Slavic past.

L., Hawrelko, R., Senych.

Furnished by Author

One of the fruits of their labours is the two-volume set: "Building and Working Together, A Study of the Thorhild Area." John Hawrelko's fine "Italian hand" pervades the cover and the text. His beautifully stippled illustrations add uniqueness not found in other works. Mike Senych organized the text, which contains his autobiography for posterity. This work is truly a treasure trove for the present generation, as well as for those to come, in understanding their heritage. It provides insights into understanding the culture of those days before and during the boom of 1948.

John's many years as a teacher have not been forgotten. Corrine Rault:[4] "I remember him reading stories, poems, and allegories ... he translated and explained ... great class discussions encouraged us to

4 Nee Dziwenka, born in Thorhild and former pupil of both Hawrelko and Senych.

answer our own questions ... he provided a challenge." Then, referring to Mr. Senych, "What they did for us ... how they opened our eyes to opportunities ... despite the authority of the educator ... very approachable ... the best parts of our lives ... the fun years." Having earned grudging acceptance by the Anglocentric world, they are now the envy of many of their western European counterparts. It was not enough just to excel; their determined drive made it possible for them to reach this level of esteem.

John Hawrelko exhibits a very strong sense of ethnicity, a sort of resistance, which continues by reliving his Ukrainian heritage. The author's three meetings with him were stimulating in that we were probing each others' minds: the Anglocentric world versus the Slavic introverted mind, each analyzing the other's differing perceptions. The author's role was not only that of an interviewer researching Redwater but also as a provider of information about the workings of the oil field, and a unique, lasting rapport has been achieved.

Stories from Hawrelko's past serve to illustrate his strong links with his people's customs. First, his father's misfortunes during the European war of 1914-1918 (WWI).

> ...My father had come with what is called the second wave of Eastern European settlers to Canada.. before WWI began. In January of 1914, he went back to the Old Country because his father was ill. Grandfather wanted to settle the estate before he passed on. My father was the eldest son. In the Ukrainian family the eldest son is usually the one that is groomed to take over the property. As it turned out, my father came there in January and, of course, the Great War began that year. And because he had been in the military for two years prior, the authorities wouldn't let him go back to Canada. So he was conscripted immediately and sent to the Front. He subsequently became a prisoner of war and was eventually interned in Siberia...

> ...He was a POW, and he wasn't able to return to his village until the spring of 1918, after the collapse of the Russian Empire. And for all those years my mother had no idea that he was alive. He was so disillusioned with what war had done to the villages and to people, he said he wasn't going to stay around very much longer...

> ...The real problem was that the authorities in Canada, prior to 1913, really hadn't an idea who these people were. They were variously called Ruthenians.. White Russians.. Ukrainians.. Austro-Hungarian citizens; they were also quite correctly called Galicians. No one really knew what moniker to hang on these people. As a consequence, my father had no "papers." When he immigrated in 1923, he became a

naturalized Canadian. It took months before he became a citizen of Canada.

...So in July 1926, my mother came to Canada. She was appalled by the sand hills, swamp and mosquitoes, having left behind the Carpathian foothills with their cherry blossoms and apple and pear trees ... She told me that she cried for a year before she settled down...

John describes European marriages.

...My oldest sister, who is now deceased, was married at age 15 in the Old Country. Marriage for girls at that time occurred any time between the ages of 14 and 16. Households that had girls beyond age 16 were quite disadvantaged: What were they to do with all those girls in one house who were not married..?

Hawrelko's comprehensive knowledge of both Europe and Canada are illustrated.

...The settlement around Bruderheim predated settlement in the area around Redwater, Opal and Eastgate by 10 or 15 years.. Some of the people who came from the area of Galicia in the Old Country were sort of indentured.. Perhaps that's not the right word, but they certainly seemed to be assigned to German families in the Bruderheim area. There was a considerable Moravian settlement and, because some of the Ukrainians had some facility with the German language, they worked for them until they finally were able to establish themselves on land in the Lamont-Star-Edna area...

I know one of the real drawing cards for this area was the fact that the land was forested with poplar. The Ukrainian settler had been deprived of any kind of wooded plots in the Old Country. Any forested land belonged largely to the Polish overlords, who controlled access to the forests for firewood and building material. So when the Ukrainians came to Canada and saw the poplar forests, they were overjoyed with the possibility of owning 160 acres of forest. They weren't exactly agronomists who could understand that poplar forests don't necessarily indicate good soil.. But they were, nonetheless, delighted with the miles and miles of forest. That is certainly one of the reasons why they chose to settle in Lamont and eventually over in the Redwater area. But the (North) Saskatchewan River was a real barrier to the movement of people from the southeast to the northwest. The Edna-Star area was well established before there was any penetration of the poplared forests north and west of the River...

Hawrelko recalls how he pulled himself up by his bootstraps.

...Summer school was the way to go, because there wasn't any money for a full session at university. It was a different story for my Master's..

> But, the academic pursuits were no different, I guess, from the other: a matter of studying and teaching school and working on the farm and things like that.. a kind of gruelling drive by degrees. I taught school in the area for a good number of years."

When one thinks of Mike Senych, the mind conjures up a whirling dervish - constantly on the go from one meeting to the next - never still. You have to contact him before 8:00 a.m. if you want to talk to him. And that boundless energy is indicative of how he served as an MLA in the Manning-Strom government with three months' leave of absence while he taught school serving as principal.

Michael was born in Corbin, in the Crowsnest Pass, to John and Anne Senych (see pp 188, Memories; also pp 898-902, Thorhild book). Mike's mother died shortly after his birth, and father, John, moved back to Redwater. John married Eunice Melnyk, daughter of George and Marie Melnyk of Opal. There were four children from the second marriage.

Mike has a Masters of Education from U of A. He taught in 1948 in Coronation and didn't return to Redwater until March 1950. "The discovery well was only a mile east of where we lived. No one knew what was going on. When I returned and got off the bus, I was lost - in my own community! I didn't know where I was. The boom had hit."

Mike describes the Redwater brick schoolhouse, where he taught Grades III and IV until September 1956, a storage room converted into a classroom and the coal room in the basement converted to quarters where the teacher could stay. The classes were jammed. He recalls the high number of transients. One little girl cried in class - unsettled, she couldn't find a friend ... this was her eleventh school. Along with overcrowding, the School Board couldn't do any planning.

His father's property was Sec. 31 (Twp 57, Rge 21 4M), a Western Leaseholds tract. He was asked by Petrofina, who took over that parcel, "Would you be interested in personnel work?" He declined. He transferred to become principal of the Thorhild school. He was also an MLA for Redwater from 1963-71. Senych is still active, serving on community committees.

HRYNCHUK FAMILY

Fred and Sophia Hrynchuk immigrated to Canada after their marriage. They purchased the "home place," NW/35-57-22 W4M from the CPR for $28/acre in 1919, complete with stones which took two weeks each spring to "harvest". The couple also acquired the NE/25-57-22 W4M.

By 1935, the family was complete: five boys and one girl. But disaster struck that year when Walter was only ten years old; his father was kicked by a horse. According to Mike, the eldest, it took them almost four hours to get to Radway hospital, where Fred died shortly after. As with so many country folk, there was no will, and the estate was administered by the Public Trustee until the youngest, John, had reached his majority in 1952.

"Mother kept us in line. She made sure that every spare cent went to pay off the mortgages," and this was accomplished in 1940. That same year also marked Mike's marriage, but his bride could not get along with Sophia, so the couple moved to a dilapidated shack on the NE/25. By 1943, they had moved to their own quarter. In 1952, Sophia gave the NW-35 to John and Peter. Walter had already been given the NE/25.

Mike's only contact with the oil patch was in 1948, when he was approached by a Vancouver firm on their false premise that he had the minerals. A trip into the Land Titles office settled it: the CPR had (has) the minerals.

Walter Hrynchuk's melancholy, a sort of "weltschmerz," never far from the surface, manifests itself openly when he relives family troubles. Disputes over his father's estate rankle in Walter's telling of them. A separation involving one of his children saddens him. Sophia passed away in 1978 at the age of 84. She was buried on May 29, Walter's birthday - a fact which he continues to find galling. Fortunately, his wife, Vera (nee Wosnack) and his bachelor son, Larry, provide support.

Walter left school in 1940 to help his mother on the home place, NW/35-57-22, and provide for his two younger brothers, John and Peter. Walter and Vera were married in July 1948, just before the well came in. His mother had asked him what parcel he wanted. "I'll take this one (NE/25-57-22) and I'll start from scratch ... because there was no house ... nothing." They later built a fine dwelling which they occupy today.

The Hrynchuks, January 6, 1955. L to R, John, Mike, Peter, Bill. Seated: Nancy, Sophia (matriarch), Walter.

Furnished by J. Hrynchuk

Walter recounts the chain of events respecting the dilapidated old shack on his present property. "During the Depression years, it was a home for people that had no place to go, no work. They would get $5 a month relief from the Government, and the farmer paid $5. And you got him to do some work -threshing on the field, or something like that, for that $5 a month that you gave him."

John was a rawboned youth of 18, just graduated from Grade XII, when he became one of the first hands to be hired by Anglo-Canadian Oil, which had just purchased E/36 and wanted to get it drilled fast. Let John describe his first day:

> ...My brother, Walter, and I drove up the first part of December, because I started on December 10th with Anglo-Canadian, and asked for the Foreman. He said he wants both of us to come to work. As it turned out, the next morning it was 40 below, and Walter said, "It's nice and warm in the house. I'm not goin' no place." So I went to a friend's place and said, "Hey, Walter doesn't want to come. So you want to come with me and go to work?" And he says, "It's warm in my house, too. Heck if I'm goin..."

So John soloed it: "My first day, picking dirt with a pick."

Production foreman George Thurber (from Turner Valley) asked John, "You want to work or you want a job?"

> ...I says, "I'll work." So I worked for one month on the construction ... And then on January 8th or 9th, our first well came in, and that's when I started shift work as a battery operator...

To give an idea of the high level of production, John states there were six lines coming off the tanks, meaning that six trucks could be filled at one time. No wonder they were able to produce thousands of barrels a day. There were no dikes, and it was inevitable that there would be spills. "There were no allowables then, so you could take all you could truck away." John reckons that the half section was paid out by April 1949.

It wasn't long before Hrynchuk was elevated to foreman, and by that time the companies had acquired the NE/10-58-22 at a crown reserve sale. He progressed and was put in charge of the Redwater operations, superintending twelve wells and reporting to Bob Lord at Camrose and Don Storey at Woodbend. Despite the fact that Federated (Home) had bought some crown parcels down by the river, there was no move made to combine lease operations.

John recalls the abrupt takeover by Triad (later BP, and still later Talisman) on January 1, 1961. All of a sudden, two individuals showed up on the 31st of December and said, "We're coming tomorrow morning at 8 o'clock to do the cutoffs." John was going to a party:

> ...And we were out 'til about 3 in the morning. When I came out to the lease about 8:15, these guys were taking stock of the oil in the tanks and all the equipment ... They were some kind of executive ... they both had suits on...

Hrynchuk reported this to Bob Lord, who said, "John, why the hell did you let 'em in?"

> I said, "Well, what could I say?" He says, "Well, you coulda chased 'em out. They had no rights.. no clearance with the Calgary office whatsoever to go out and do this." I said, "They came.. so this is it." And that was when they took stock and said, "Hey, we bought Home Oil."

Shell bought out Anglo Canadian in 1962 and John was moved to Edmonton, the Miller Building on 105th Street. He confirms the fact that most of the Anglo people were shoved aside - a desk with no paper on it. "Turk" Terlecki was one of the casualties.

Hrynchuk's ability was soon recognized by Shell. Hrynchuk described their Worsley spread in northwestern Alberta.

...And there was one stinking oil well out there, and they didn't want to look after it. They tried to get the gas plant to look after that one well ... The pumping unit there on the well had a high gas/oil ratio, so the guys were drawing their own charts. And I guess they got caught, and the well was shut in for so many months because of the penalty...

The author wondered how one would get those jiggles on the charts, and John laughed, "Somebody had a shaky hand from the night before."

Hrynchuk was told in March, 1965, "You're the new Maintenance Foreman for House Mountain!"

...I said, "Why me? I never had drilling or completion experience, or workovers." "Time you had some." So, I went up to House Mountain and left Rose in Edmonton."

John and Rose Hrynchuk in their back yard, Redwater.

Furnished by Author

...They asked me to move to Swan Hills ... and I knew I was going to quit ... The girl in HR Department said, "John, where were you hired?" I said, "In Redwater." She says, "Don't tell anybody, but they've got to pay for your move." So, two months later, I sent my bill in. They paid it...

He recalls his confrontation with Mac McLachlin when he brought in his brief case and the keys to the company truck.

> ...McLachlin asked, "Will you reconsider?" And I said, "Such as what?" He says, "Well, more money." I said, "Hey, you raised me $11 to go into the bush, and you told me I was overpaid ... Don't talk money to me now. My family's more important."

> ...I retired from the Imperial Fertilizer plant effective December 31, 1990, after 22 years. Imperial offered me a good package ... would have been silly if I wouldn't have taken it ... age 60 ... Imperial was good to me...

He recalls Dec 17th of 1990:

> I didn't have to get up and drive through the storm ... I missed the people, but not the job...

> ...I think, overall, with no university education and growing up on the farm, I've done pretty well, along with getting married to Rose and raising a family of four. And it takes the support of the family to make a livelihood. I think we worked it out well -we've got a nice family...

Rose's maiden name was Wispinsky, and her roots were in Andrew. Her grandparents had come from the old country in the late 1890s. She recalls her Dad's cafés in Edmonton, successful operations, busy all the time. John and Rose met in Edmonton in 1950 and were married there August 9, 1952. They have raised three daughters and a son. Rose's youthful appearance belies the fact that she has four grandchildren.

ANNE KEY - TEACHER, MOTHER AND HISTORIAN

Anne Key (nee Yachimec) takes first place in a list of those whose roots are in that faraway province, Galicia. There never had been a time, it seemed, when this area was its own master - Poles, Prussians and Swedes had overrun these lands. The Austro-Hungarian Empire had control up until World War I, resulting in many immigrants from the Ukraine being branded as enemy aliens during the two World Wars. Indeed, some were interned. "I knew what it was like to be in that category ... I was fingerprinted and the whole bit," as she recounted the stories of her father, Michael, who arrived in Alberta in 1907 "with ten dollars, ten fingers, and no tongue."

Anne in her family room.
Furnished by author

The homestead was located four miles from Egremont : N.W. ¼ Sec. 34, Twp. 58, Range 22 West of the 4th Meridian. It consisted of 155 acres of bush. There were no roads or railways when the settlers came. The railway came through in 1914, and the roads improved very slowly. Old Indian trails were used for a long time.

The family endured the usual hardships so well known to those early pioneers. Michael Yachimec had two oxen which he used for three years to break the land with a walking plow. He also used them to make trips to Edmonton for supplies and to sell farm produce. Eventually, horses were obtained. The hard work of clearing away the heavy bush and trees eventually paid off. Like most farmers, Michael soon had over 100 acres available for crops of wheat, oats, and barley. His farm operation prospered, and in 1925 he was driving a Model T Ford. By 1928, he had acquired a tractor and other power machinery.

A farmer in pioneer days had to be a jack-of-all-trades, and Anne's father was a good mechanic, blacksmith and carpenter. An innovator, he strove to mechanize and automate farm operations as much as possible. He had a natural flair for business, earning extra income with equipment set up to do custom grain cleaning, crushing, wood sawing, threshing, and lumber sawing for others. While in Thorhild, Michael was active in church and other community projects. He was instrumental in the organizing and building of the Ukrainian-Orthodox church in Thorhild.

Anne's mother, Pelagia Bushko, who came from the same village (Staryawa, near the city of Lvov), travelled to Canada to meet her bridegroom via New York City. "...She came to Vegreville in a big-brimmed hat, with an umbrella and a wasp waist ... They were married in 1912."

Anne then speaks of herself.

> ...I was born in 1923 on the farm ... I started at age seven in Ingleside school. We walked the two miles ... 60 below meant nothing ... we learned the rule the first day that if you speak Ukrainian you get the strap. And I couldn't speak English. So I told Mother, and she said, "Keep your mouth shut and your ears open"... Teachers were Ukrainian, but they weren't allowed to speak Ukrainian ... What makes you think you can do the classes when you can't even speak the language? I said to myself. "You watch," and before too long I was leading the class...

By 1942, still in her teens, Anne had qualified as a teacher and she had several assignments - to be broken by a three-month hitch, along with 100 others at a munitions plant near Ajax, Ontario.

When Redwater crude was discovered (Oct. 1948), she left teaching and moved back from Sexsmith with her then-husband. She looked after her two children living in a garage built by her father. But she later returned to the classroom, her true love.

Despite her infirmity, Anne directs traffic with innumerable phone calls from her home in the eastern outskirts of town. She continues to provide names and advice to the author, and applies the gentle but persuasive style she developed in her teaching days. No wonder she was honoured with a life membership in the Alberta Teachers Association!

John Hawrelko, retired educator, credits Anne with gathering stories for the 1972 book, "Memories."[5] She wanted her pupils to improve their writing skills, so she sent them out to interview old-timers. What a contribution, both in the past and to the future!

5 See Appendix XX for Author's 1992 book review of "Memories," published 1972.

THE KIMURAS

To get some background on the little known Japanese community, the author interviewed Jim and Chizuko Kimura. They live in a two-storey white house just off Highway 28 and west of Kimura Lake, recently named by the Alberta Cultural Historic Sites Division.

Our story starts in 1906 when Jim's father, Toyomatsu Kimura, emigrated to Canada. Menial labour was the order of the day for him and his countrymen. It was while he was in lumber camp that he applied for his citizenship papers. Returning to Japan to marry, he settled in Edmonton before WWI and was successful in building up businesses: two barber shops and a rooming house. The regulations of the day hampered his ambitions and he turned his attention to acquiring acreage east and north of Edmonton. At the height of his career in 1928, Kimura owned nine quarter sections. When the Depression hit, not only did our hero lose most of his land, his farm machinery was seized. Jim recalls, "My mother and I and a couple of hired men were at home, but we were helpless ... I saw how these firms had no mercy on the farmers." Hard times were reflected also in the construction of the two-storey house which the Kimuras occupy. It was impossible to finish the dwelling as Toyomatsu would have wanted because he ran out of money and had to dismiss his cabinet maker.

Jim recalls his Dad's passing at the age of 91 and his mother's at 87, eliciting the comment that longevity was in their genes.

Chizuko Matsuno's family had emigrated from Japan in 1909. They found their way to Raymond and developed a prosperous ranch in the Milk River Ridge country to the south.

The huge iron stove in the kitchen had been converted from wood and coal to Co-op gas, and when the author visited them it was the main source of cozy heat. Equally important, it was keeping Chizuko's sourdough batter warm.

Chizuko combines her prowess in prize winning cooking with a composure and serenity that are due in no small way to her devotion to the Buddhist faith, a belief that has carried her through darker days.

When the boom hit, Jim went to work as production hand for Pacific Pete. When the Redwater job ended, Pacific wanted Jim to go to Drayton Valley but, because his father was getting too old to farm, Jim turned down the transfer. Instead, he went with Tartan

Jim and Chizuko Kimura.
Furnished by Author

Construction (Bert McDonald) in Redwater. When he left Tartan, he formed a company (cleaning up and rebuilding batteries) but it soon folded.

Summing up, Jim states

> ...I am very fortunate to have lived in this area. The people were all very friendly, even during the war years and the Depression. We were all poor. I enjoyed sports ... baseball ... had great times ... life was very good to me. Most of all, I was fortunate to have met Chizuko and raised three good children...

Chizuko added

> ...We're grateful for the friendship in the Redwater area ... always included in the activities there. We were blessed with good health, children ... certainly living in a wonderful era. I think our parents enjoyed that most...

KIYOOKA: ARTIST IN THE OIL PATCH

The story of George Kiyooka, a talented artist, exemplifies the desire of many Japanese, among them his father, to immigrate to Canada from crowded, oppressive conditions. Most Japanese stayed on the coast, but some found their way to Edmonton and to Raymond. Barely tolerated by the Anglos, they prospered. The next step was to become farmers, settling as early as the 1920s in the area west of Redwater. But it would not be until after WWII (1947) that our hero and his bride would move in to the well-established but little-known Japanese colony in Opal.

George was born in 1918 in Victoria, B.C., of Japanese parents. When he was two years old he was sent to Japan to live with his grandparents because both of his parents had to work. George's primary education was in Japanese, so he had to start schooling in Grade I to learn English (Stanley Jones School, Calgary) when he returned to Canada.

George's Dad had been running a grocery store, but by the time George arrived in Calgary in 1931, the business had failed. He then found work with his father at the Empress Hotel (217 - 6 Av SW, now demolished, then owned by A. E. Cross brewing interests). Shortly after Pearl Harbour (December 7, 1941) he and his Dad were told they no longer had jobs and were treated as aliens even though George was a bona fide native-born Canadian citizen.

Two years of commercial art at SAIT followed to develop his latent ability to draw and paint.

Rod Morris, Imperial Oil wellsite geologist, had moved into a rundown two-storey house in Opal with his wife, Jean. He befriended George, who was now in Opal and looking for a job. Rod drove him in to the Redwater office and had him hired as a sample washer (in the chicken coop). George recalls Rod "never was really at home at Redwater ... he was a different type of a guy."

When Reg Shatford (George describes Reg as an academic more than a manager) took over from Rod he noted the artistic printing on the shipping labels and immediately switched Kiyooka to drafting subsurface geological maps. George tired of the bureaucratic life in the office and asked to be transferred to the Production Department as a battery operator in 1953.

Redwater was too far a commute from Edmonton for George's children, who were soon to be university students. In 1956, he heard

Igor, George, and Bob Kay.
Furnished by S. Bleakley

that a packaging and blending plant for lube oil would be opened up in Strathcona by Imperial Oil. Once again he asked for, and was given, a transfer.

The family moved to 998 Conifer Street, Sherwood Park. A second storey was added to provide room for them all. The 5 per cent 25-year mortgage was discharged in 1981. Retired in 1983, George describes his family triumphs tinged with tragedy--deaths--one son of leukemia several years ago and his wife in 1991. Kiyooka lives alone now, filling his days by transmuting his artistic gifts onto canvas with exquisite renditions, especially of birds. His diction reflects his childhood upbringing in Japan but he is a true Canadian waiting for the golf season to start each spring.

LAKUSTA CLAN

A steely resolve, obvious in her 1926 snapshot, is still evident in Helen Lakusta's deportment. Although arthritis has partly immobilized Mike's 85-year-old widow, she still exhibits the independence which has carried her through a long life. Helen still bakes doughnuts, marks the sale items in the newspaper, and orders her groceries every weekend through her children. This enables her to stay in her home. And her children are proudly displayed on the walls of her home in northeast Edmonton with good reason, all of them - and their children - successes in their own way.

The first record of the Lakusta family is having homesteaded in 1899, filing on adjacent quarters of crown land (surface only) by brothers Fedro and John: NW35-56-21 and SW2-57-21, respectively, just north of the river.

Mike Lakusta's parents had come from the old country around 1899, and he was born in the dead of winter - January 17, 1900. Ed Lakusta, Helen and Mike's eldest child, recalls stories told him about his grandparents.

> ...They settled in Andrew. Their first home was a hole in the side of a hill with sod and trees thrown over it. They had heat inside, and they lived that way for a year before they built a house ... They got through the winter by picking enough berries and living off the game and fish in the Redwater Creek...

Ed's grandfather had heard somewhere that there was a railway being built through Fort Macleod. So he left his bride and young children on the homestead and walked all the way to Macleod to get hired on.

> ...He worked there all summer and saved enough money to buy a wagon, a set of harness, two horses, and ... something like $60 in his pocket. He came home in the fall.. realized he had no feed for the horses, so he sold them ... When spring came.. he bought the horses back for a discount - because they were thinner - and he started breaking the land. Grandfather moved from Andrew to Cookville ... and he would walk to Edmonton, buy his groceries and supplies for the winter, cut logs at the river bank, build a raft, load his stuff on the raft, and float home to Cookville. He would dock the raft over on the left side of the North Saskatchewan river and then walk home, get his team and bring his supplies home..

Harking back to Ed's father's birthplace, "My grandmother would.. sprinkle water on the walls of the dugout to keep them intact and minimize the dust."

Helen's parents, the Dowhaniuks, settled near Waugh, eight miles north of Bon Accord in 1903. Helen was born May 16, 1908. When asked why her parents had immigrated from the Ukraine. That's what they were going for." Describing her father's farm, "There was no building ... Father had an accident. He fell on a stump when he was building a granary ... In three days, he was gone. That was 1916. He was 43 years old"

Helen describes the death of her mother in 1925,

> ...Double pneumonia at the General Hospital in Edmonton." She and her two brothers, now virtually orphans, were well looked after, "The neigbours were good to us, because we were good kids. Wherever we'd go, they give us something to eat ... keep us, like orphans, the three of us...

When asked about her life on the farm before she got married, "We were doing everything - making hay, hauling ... I was building a stack ... crawled down ... the stack tipped over, followed me ... I didn't have it the right way." Helen was one of the earliest females in the area to drive a car. When she married in 1927, she left her '23 Chevrolet with her two brothers on the homesite.

Asked how she met Mike Lakusta,

> ...At the Opal dances ... We knew each other for about two years. I wasn't ready to get married at seventeen ... but at nineteen, I was old girl already. Mike was the kind to marry ... he was a good husband- 27 - he wasn't a boy. He had a farm and horses. That's what you married for, those days.. And he was a hard worker. And he said, "My kids aren't going to have to work like me.. they have to be in white shirts and ties.. They're not going to live on the farm and work the way I do"...

There was no honeymoon: "The next day, we went to haul hay ... July 14 ... Everything has to be done just so ... A time for everything."

Helen describes her children chronologically: the firstborn, Ed (1930), a daughter, Mary in 1932, Russell in 1933, and then in 1938 - Vic, the baby.

It wasn't long after Imperial's discovery on Cook's farm in 1948 that Pacific Petroleum, who had purchased the oil and gas rights under Lakusta's property (S/10-58-22 W4M), leased the surface from Mike.

Helen beside her 1926 Chev
Courtesy of H. Lakusta

Helen, 1993
Furnished by Author

Helen described her husband's worsening heart condition, "The doctor said, 'Don't touch nothing.' And no telephone, I didn't want to stay there." Mike passed on in 1970, after living in the city five years.

Comparing her current activities to the old days, "We used to buy 10¢ socks for my husband.. and now they pay $14! ... I'm just getting the Old Age cheque ... but some of them are getting spousal benefits ... And you know what they do? They play Bingo every night ... or drink ... and kill each other."

Bingo is Helen's main recreation. "I used to play twelve, nine.cards. Now I'm cut down to six. I like to laugh. I don't get mad that I can't win.. And I got a good friend who takes me to the Bingo games."

Ed describes his education,

> ...Grades I to XI at Opal ... then University high school in Edmonton, "And I graduated in June of 1949 ... It was kind of dramatic for a farm

boy going to Edmonton to school. Things were pretty tough, and there wasn't much money around ... I had a little room in the basement of a house about two miles from University High School. The window was my refrigerator and a two-burner hot plate was my stove ... But it was exciting ... Education, to my father, was very important, because all he had was Grade III ... a self-taught man,[6] so he really pushed us to go to school. The routine was, I'd catch a bus on Sunday night to go to Edmonton ... I'd get a cream cheque ... about $4.35 and a little suitcase of groceries that might last me for three days, because that window wasn't a very effective refrigerator. And I had to sparingly use that $4.35 to buy my groceries and save $2.35 for a round trip fare to the farm on the weekend ... And I remember bread was 11¢ a loaf...

Asked his perception of Leduc,

...We saw the smoke when Atlantic No.3 caught fire- we could see it south of the school I attended. The University High School, incidentally, was the old Normal School, and they used the students to train and develop their graduating teachers; the students there were guinea pigs. So, we'd watch the smoke.. for a long time. There was a lot of media coverage ... so we had a good perception of what was going on...

When oil was struck at Redwater, Ed got a job with Pacific Petroleums.

...After completing high school, I didn't know what I wanted to do ... Pacific were right on our property,[7] so we got to know a lot of their people ... I remember Buster Boyd and Bert Frew, Lyle Caspell.. Scotty Tosh ... who were.. very visible. I guess the guy who hired me was Buster Boyd ... My starting salary was $1.08 an hour ... in three months I was foreman of the whole crew. And in nine months, I was kinda running their maintenance operations. I think my salary got up to about $1.35 an hour ... I remember - living on the farm[8] - you not only worked for Pacific, when you got home you worked on the tractor 'til midnight ... I had free room and board...

Gulf came along - on one of the quarters we acquired, the NW16-58-22 - the M. J. Lakusta Lease ... I still own the land. And that's where I got my introduction to Gulf. You know, I liked the way they did business. I worked for Gulf 'til about September 1952. By then, switching became a little boring.. I could go up there in the morning and throw a rock at the tank, and I knew how much oil was in it. It

6 Helen states that Mike would read the paper every morning.

7 S/10-58-22 W4M.

8 Both refer to the South Half

got so I didn't need to gauge 'em; I knew exactly what was in 'em ... I talked to Gulf and pursued ... going back to university and getting leave of absence. And they were gracious enough to see me through ... employed me during the summers in between terms to generate enough cash to go back to university again...

"I started in Mount Royal in 1952, went one year, and then I went to U of Oklahoma for two years ... there were 27 Canadian students there that year. My total time got counted for service purposes. And then ... I had employment during the summers And the year I roughnecked, I ... walked away with a net of about $3500. I worked 16s (two 8-hour shifts, back-to-back) every second day filling in for people taking vacations. And I was just a young fellow - 16s were nothing for men ... I was able to go back to school and finance myself...

Ed describes Kelly Gibson, fondly referred to as "Croc" in Stettler.

...I didn't get introduced to Kelly 'til the year I was roughnecking. I saw more of him when we were well testing ... Kelly was a tough taskmaster, but those were tough days, and he came up in the hard school of knocks ... he got it done. That was the day when you were told you were transferred, and you were also told that your next cheque was going to be at the new location. And if you didn't want to move, you didn't get your cheque...

Gulf had a very good program ... to develop engineers. In fact, you'll find that Gulf-trained engineers ... did very well. That's true for all the majors ... the oil industry was growing very rapidly, they had a large demand for people ... they recognized the need - that developing people was very important. Naturally, they couldn't hold them all 'cause there were so many opportunities ... so much growth in the industry ... I think they did a great job for the country, for the province ... and the people...

We had a large number of moves over the period ... Pincher Creek, Stettler, Edmonton, Calgary ... spent some time in Toronto in the Marketing Department, then in the U.S. in their downstream operations...

It was at this juncture that Petro-Canada came along and offered him the position of Chief Operating Officer, a thankless job from which he retired in 1991.

When asked about abandonment of old properties,

...I know some of the companies are setting aside funds for old fields and wells. That's going to be a significant cost ... The worst is going to be those old tank batteries and flare pits ... that's going to take a long time to clean up. And then any major spills, and there's been a lot of them ... and as I mentioned earlier, as these fields matured, and water production increased percentage-wise ... if you get a spill or a failure in a flow line between a well and a battery, you're not getting just an oil spill, you're getting a water spill. When these fields are depleted and become uneconomic, the sites will have to be restored to their original use - to agriculture...

Lakusta Family L to R, Russell, Helen, Ed, Mary, Mike and Victor, circa 1965
Courtesy of H. Lakusta

ED LAMBERT

Mail courier, jug hustler and maintenance foreman. It was a bright, sunny, late Fall day when I visited Ed and Barbara in their Redwater home south of the tracks. Three cats, all of them different breeds, dominated the household - as cats usually do. Up on the shelf was a huge fish tank with what looked like a piranha ready to attack.

Ed and Barbara reviewing one of their many photo albums.

The first thing I had to do was slow Eddie's machine-gun manner of speaking. The second thing was to try to understand his unbelievable stories. (There's always an element of truth in them.)

We started off with his famous father, Thomas Scott Lambert, whose roots were in Fort Vermilion. He had come to Redwater to marry Ed's mother, a Soucy. Scott was well known for milk deliveries from their house where the CIBC Branch now stands on Main Street. He was also the mail courier for the outlying Sub-Post Offices, Val Soucy and Eastgate (once-a-week deliveries). Scott was noted for his reliability and John Myroon remembers this trait. Ed tells a story about his father, who used a closed cutter in the winter pulled by one horse. It contained a coal stove. One time, the horse ran away through town and upset the cutter. "Dad was covered with hot coals and ashes, with some burns, unable to extricate himself. Finally, the neighbours rescued him."

Ed tells a similar story about his own mail trips. This was in a blizzard when the cutter got too heavy, and he unhitched the horse. John Myroon wanted to keep him at Val Soucy for the night, but Ed insisted on walking back with the horse, jumping through drifts, then resting. They ended up at the Brown house (now a deserted dwelling on Anne Key's property), where he was met by his Dad at eight o'clock that night.

Ed attended Ufford School, about a mile and a half away, and his education ceased during Grade XI. He recalls little of the early days of the oil boom because he was working part-time for his Dad and for the CNR.

Sometime in 1948 he showed up at Malowany's Garage, where Heiland Exploration kept their trucks. Ed recalls being hired by Warren Gunn, Party Chief from Shreveport, Louisiana with his southern drawl. The other person on the Party was Bill McPhee. Ed laid out geophones, 24 of them, all wired together. This must have been one of the crews that Ray Walters hired to do detail shooting prior to the drilling of Imperial Redwater No.1.

There was a short interlude with the CPR at the Strathcona Roundhouse in south Edmonton. This is where he acquired his hobby of collecting and/or building model locomotives.

Lambert returned to Redwater to work for Tartan Transport - run by Bert McDonald of pugilistic fame. The outfit had about eight trucks which they used for general oilfield work. He remembers being told by his boss to drive over the North Saskatchewan River despite the fact that the ice was black. He put the truck into third gear full throttle but "I felt myself going, and the truck broke through the ice and into about eight feet of water." Typically, when Ed got back to Redwater, his foreman told Bert , "Eddie put the truck into the river."

A kaleidoscope of stories follow: the first sand-frac with Halliburton in Canada, battery operator for B-A, cleaning out waxed tanks - all of which culminated in Ed getting a job with Esso Chemicals at their Fertilizer Plant. He ended his working career after 15 years as Maintenance Foreman, retiring in 1990, and now spends most of his time collecting information on steam locomotives.

MARY MALOWANY: MATRIARCH WITH A MISSION

Mary Wacko was born in Edmonton. She is the widow of Walter Malowany,[9] who was prominent in both business and the Redwater Town Council.

Fashion plate of 1936: Mary Wacko standing on the train platform at Myrnam..
Furnished by M. Malowany

Our interview started with a display of ancient documents: "Our records show that Stefan Wacko, aged 26, married, arrived at the Port of Quebec on June 1, 1910 aboard the S.S. Mount Royal. This man was born in Austria." Mary remembers her father telling her about being in the Austrian Army. He could read and write in Polish and

9 Lauder Nowers, then head of Imperial Oil Personnel at Redwater: "If we had a problem, we went to Walter, and you could almost say it was solved by the time you explained your needs to him." For instance, the Royal Bank urgently needed space, and Malowany made room in his Garage for Bert Smith, the Manager.

Ukrainian, and after he had arrived in Canada in 1910 he went to night school to learn English. With this ability to communicate, Wacko obtained a job with CNR as a 'section man' in south Edmonton.

Mary went on, "My mother was illiterate. She died that way. She didn't want to learn." This was part of the reason that Mary's education stopped at Grade X,

> ...Then I quit because my mother said, "Well, the girls - they don't need education." I was very saddened by it, because I was very good in school ... but she insisted that they weren't financially able to get books for me, and things like that ... You know, in those days, 'No' meant no. That was 1930...

Then, talking of her experiences trying to get a job, "...You could get a job doing housekeeping for $5 a month or, if they were wealthy, you'd get $10."

Stefan's sisters had come over from the Ukraine, "...They wanted land ... a homestead. So we helped them ... set them up on farms ... but they always came to visit, because my Dad had a cheque every two weeks - $100 every two weeks. So, they would want to come and borrow." (And the inference is that Wacko, Sr. willingly helped them.)

Mary describes how she and Walter met. "We started to correspond, and he'd come in - oh, once every three months or so. And I'd meet him uptown by the Eaton's store, and we'd go out for lunch. Then he'd go home, and I'd go home."

Mary recalls Walter's job with the Gillespie Grain Company when they got married in 1937 and his salary of $100 a month. The house which went with the elevator was part of Walter's salary. As with Wacko's sisters, the father-in-law came through with finances to get the couple started on a quarter section.

By 1942, Walter had rented an old garage and was handling Cockshutt machinery plus the Pontiac-Buick dealership. Mary was not happy with the Gillespie house, "...I was expecting another child, and I said, 'You know what? Let's buy this house, tear it down, and build a better home.' And he said 'Okay'."

When the boom hit, business really picked up. Mrs. Malowany remembers helping out in the garage, doing bookkeeping.

> ...There were a lot of U.S. outfits ... There was a lot of confusion - they would run up a bill, and then you couldn't trace them. You know, there was a lot of that. Yes, they'd skip. Or else they'd pay a little bit and say, "We'll pay later."... He (Walter) lost a lot ... They got the service and they didn't care to pay. But everybody thought, 'Oh my,

> Redwater's got the oil and it's going to be so prosperous.' But it wasn't - not for my husband, although he was a good businessman...

Ever proud of her children, especially since she was denied an education, Mary outlined the career of her son, Peter.[10] He started off by winning a Grade XII Hotelmen's Scholarship of $500. He graduated in engineering from University of Alberta in 1959. A successful international career with Chevron ended when he decided not to transfer to the Middle East. He found no difficulty in relocating. Married with children, at age 54 he is now Director, International Development, with Norcen Energy.

Mary next describes her daughter, Marge, married to an oil executive and living in Calgary. She won a Teacher's Certificate as well as a Bachelor of Education degree. The younger daughter, Virginia, graduated from University of Alberta with a degree in Art. She married Paul Sharek and has two boys.

> ...I wanted something better for my children ... because life was very difficult. Well, marriage is difficult ... I always felt it could have been different if I'd had an education...

Her husband died in December, 1990. "He was depressed, disappointed, and not too happy about his retirement."

But, looking on the bright side,

> ...I've spent a lot of happy years in this small town. I like the small town living. It's friendly and the people ... 50 years of my life here, and I enjoyed all of it...

When asked what she thought Redwater would become, she thought that it would grow, "Better system of education, better schools ... The children are better off in a small town."

10 Neal Collins, Schlumberger engineer, recalls Peter as a child running around his Dad's Garage.

JACK AND KAY MCCASKILL

When Jack McCaskill returned to Redwater, Vern Hunter was superintendent. He compares Bannantine (great for delegating) with Vern (hands on about everything). According to Don Hunter, Vern's son, the McCaskills moved in to a house behind the Hunters when they came back, and Don's brother, Gregg, played with Jack's son, Bob.[11]

Kay McCaskill fondly recalls her second sojourn in Redwater, this time in a proper company house. One of the most important social developments, Oil Wives, was started there by Dean Hunter, wife of the district superintendent, Vern.

Jack and Kay McCaskill, 1951
Furnished by J. McCaskill

Jack corroborates Doug Gamble's story about Bert Mcdonald and van der Grift. Jack remembers "Bert... he'd fight anybody..." It seems

11 Vice President, PetroCanada.

as if Vern Hunter couldn't handle them at Devon, so he sent these two up to Redwater. Jack also confirms having to drive both Bert and Van to the Radway Hospital; apparently Bert had been hit on the head with a crowbar. Such was life in those days.

An interesting story involving the McCaskills is told by Oscar Erdman, retired geologist. He had been sitting a well being drilled by Tom Brook on a farm-out from Gulf north of Radway. He was wakened one morning by a cab driver with an urgent message from his superior, 'Phone Bob Lockwood in Calgary immediately'. The reason for the upset was that, somehow, Tom Brook had acquired a piece of wireline core which had white lime in it - but no oil staining or porosity - and on this basis mistakenly thought he had hit reef. (This was also the occasion for the Grey Cup, which Tom attended; the TransCanada Air Lines fare was $129, and the plane held 21 people.)

Tom's company, New British Dominion, were trying 'to make a market', and welcomed any news which would push up the stock. Erdman remembers coring "all kinds of footage of green shale." The well was subsequently abandoned.

Oscar remembers being invited to the McCaskill's first "home" for either Christmas or New Year's dinner, very likely in the granary at Radway which they were renting . "This hospitality was the first in a long series of invitations exchanged between all of us who got our hands dirty."

ROD, JEAN AND HEC

"I am fortunate to have had two men like that ... I've had a marvelous life ... I wouldn't change anything." So reminisced twice-widowed Jean Morris-MacRae in her Varsity Estates Drive home during the author's March 1993 interview.

Harry Rodney (Rod) Morris, Edmonton native, and Jean Robinson, born in Vancouver, were sweethearts at King Edward High School. Hec MacRae was also a classmate of Rod's. In fact, he and his wife-to-be double dated with Rod and Jean. This very strong friendship was to change in 1988. After Rod's sudden death in 1983 and Hec's wife's demise following that, Hec and Jean got together. They were married only one month when the groom died suddenly in 1988. Since then Jean has carried on in her sprightly, no-nonsense, business-like way, keeping track of Rod's affairs through his company. She finds time to visit shut-ins and to read to students twice weekly.

Jean Robinson . . . sweet sixteen, standing on 600 block, East Broadway, Vancouver.
Furnished by J. Morris-MacRae

Rod received his BSc from UBC in 1943 and, as Jean says, "I got him the next day in marriage." Events moved rapidly and Rod was hired on by Imperial Oil at $65 a month plus field allowance. Rod's first job was at Maple Creek, Saskatchewan, under Fred Killer and Glenn Fox doing hand auger drilling. "Our own tent ... my furniture was orange boxes, and I had a Coleman camp stove with an oven on it. We cooked meals for the hands." Jean learned of the great hospitality of the prairie people. "Nobody had anything, but they shared it."

We next hear of the Morrises at Crossfield, where Rod sat on the wildcat in the 1945-46 winter which should have been a Mississippian oil well. This was followed by a sojourn at Tofield and Minburn, where Imperial were drilling gas wells for the proposed synthetic plant.

Rod was becoming an expert wellsitter, a profession which he would end his days doing. Leduc had already been discovered, but there was still a perceived need to drill deep tests, so it was out to Battle Lake in 1947 to a town named Winfield.

> ...This house (log cabin) that we lived in ... Rod lined the inside with paper, and you could hear the mice running between it and the logs. The rig moved east to a point just south of Leduc known as Eyot Lake in 1948 for yet another basement test (turned out to be a dry hole). "And we had our own trailer - that Rod built ... parked in Millet ... trailer court...

When asked about the next move, Redwater,

> ...I don't remember ... He would just come home and say 'We're going to move' and that would be it." Jean describes the house in Opal, about ten miles north and west of Redwater, "This big, old house there had an upstairs off the front room - just one great, big room on the main floor and then to the side there were stairs up to where there was a bedroom. Then beyond this big room, there was a doorway, and then there was a shed at the back ... Cold! You wouldn't believe it. The pot-bellied stove fired with coal liberated from the rig was the only source of heat ... One of the worst winters I remember...

Gloom turned to joy when oil was finally hit at Redwater. Jean describes Rod's reaction.

> ...Oh, delight. I think he couldn't believe it. They were just so excited. And he never came home, and never phoned, and I couldn't figure out what this was all about. And then, eventually, word began to drift that oil had been discovered ... I think, probably on the radio.. And then, eventually, he came home ... just exhausted...

> ...With the boom on, many people were hired to work with Rod. He got Hec MacRae -who had been in South American with Mining - a job. And Gordie Copland was their secretary ... and cooked food ... did

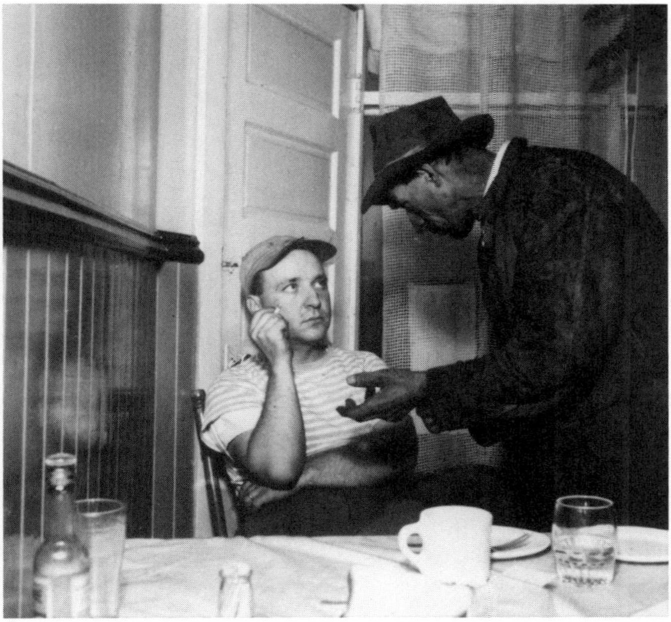

Pete Tkachuk, local farmer (having just come from church because he's wearing a white shirt, a coat, and a hat), confronting a skeptical H. R. (Rod) Morris, Imperial Oil's feisty District Geologist, in the dining room of the old Redwater Hotel, 1948 (destroyed by fire shortly after).

Courtesy of Imperial Oil Ltd.

> the typing. He was a jack of all trades ... Doug Lavoie was Rod's brother's friend... (Fred)

But Jean doesn't recall the other people. Jean has no details about Rod's sitting on wells for C. B. Barlow of Western Leaseholds.

The next move was to 2020 Bowness Road with Rod's transfer to Calgary. It was from this office nearby that, later on, he set up his consulting practice. Rod was associated with several companies: Osler, Hammond & Nanton and a combination of U.S. oil firms. Each of these outfits had its own "Jesus" department, and Rod would say, "None of them have ever been out on a well, and they're telling me what to do." His experiences would lead him to confrontations on wildcats from the Arctic to Guyana, but Rod stuck to his word by not charging more than he felt was fair despite the complaints of some of his associates about not getting his prices up. He found his real metier in teaching the art of well-sitting. Hundreds showed up for his courses, each completed with an exam.[12] His widow remembers as

12 Jean McLaws attended one of the courses while with Indian Minerals and has retained an exam paper marked meticulously by Rod.

many as 35 pupils in one class, and Rod returning home, drained. "But he had his determination and his principles. Somebody said to him, 'Why don't you charge more? Say, $2000 a pupil?' The answer was 'The oil industry has been good to me, and I want to give something back to it'."

The highlight of his international well-sitting was a Home oil well in the southernmost part of Guyana, right up against the Brazilian border. Jean recalls what went on.

> ...They had no cars. They had.. these folding bikes, and that's what they rode in to the city to report ... But, I can always remember them telling about burning the cane fields and all the snakes coming out ... This Westburne outfit were a pretty wild bunch ... Anyway, he was there probably four or five months. And he..flew home. and on the airplane were a bunch of these drillers. And one of the hostesses came up and said, "Mr. Morris, I want to talk to you ... Will you move to the opposite end of the plane?" So she said she had overheard these guys were going to get Rod. And then she told Rod, "We have arranged for you to get out of the way as fast as you can." So he rented a car and drove back to Canada. It was later learned that they were into drugs ... they figured he was going to squeal on them. Our home at 1932 - 26 Ave was given security protection until the problem had blown over.

Jean joined Rod on an African safari while waiting for his drill ship to start up. This was followed by a hole which was drilled in the Sinai Desert in 1969. "He had a contract to go to the States and teach for some big companies down there" when he died suddenly February 5, 1983. Jean describes the graphic details:

..."I don't think I feel very good.. I'll stay here for a while." This went on until about three o'clock in the afternoon, and he said, "Mum, I think you better take me to the hospital." I said, "I'll get an ambulance." "No," he said. "I want you to" ... and down on 17th Avenue ... and he said to me - and this was three o'clock in the afternoon - "You better turn on the lights. It's getting dark." And then, all of a sudden, he fell over on me ... And he died at ten 'clock that night ... aortal aneurism...

Jean picked up the pieces, finding Hec MacRae and his wife strong supporters during her bereavement. After Hec's wife died, Jean said to him, "I leaned on you ... now, it's my turn ... We're together all the time ... doing things together ... enjoy each other's company ... so, why not (get married)?" They were married March 5, 1988, but this union lasted - tragically - only one month. The circumstances of Hec's death were almost identical. It was April 5th, upstairs, and he said, "I don't think I can stand this pain. Will you get the doctor?" So I came down to phone ... I called up to Hec, "He's on the line," and I heard the phone drop, and when I went up, he was gone."

And, in her matter-of-fact way, "However ... you survive, don't you?"

BILL MUGLER

William F. Mugler was no ordinary Oklahoman. He attended Oklahoma University on an athletic scholarship - 225 lbs, always muscular, delivering 200-pound cakes of ice from horse-drawn wagons.

It was the Depression era. Bill's friend, who had been working on a seismic crew for Amerada, asked if he was interested in joining him. "And I said, 'Hell, yes. Doing what?' It turned out to be hand-drilling auger holes for shot holes ... The Labour Foreman came over ... handed me the handle of the auger ... said, 'Take this and screw it on this 3-foot length of pipe ... Get your head down and your ass up, and turn to the right.' We dug 40 feet deep, four dollars a day, no overtime."

Mugler relaxing at Canyon Meadows Golf Club.
Furnished by Author

Mugler soon advanced to work on a seismic recording truck. He dealt with the gravity meters, chemical analysis, and electrodes. He then heard of a company called British-American, with head offices in Tulsa. In reality, it was a branch office of Toronto headquarters. B-A had, since 1907, been importing their own crude from the mid-continent by tank car to supply their Canadian refineries and had built up a substantial producing company with technical expertise. Bill's first assignment with B-A was scouting and attendance at Scout Check. "We had a great, big room in the hotel ... in Shreveport. We'd all congregate there.. and exchange dope."

> ...I came into Calgary ... July 11, 1949 ... worked in the B-A office at the corner of Sixth Avenue and First Street SW, right next to our filling station. I was only to be up here for two months to set up the Land Dept. and hire help ... And then I was to go back to Tulsa." He reported to Jim Statler in Shreveport. Jim came up later...
>
> Harmon was a lawyer. And I don't know why they sent him up to Edmonton.. He was supposed to obtain right of way for pipelines.. Being a lawyer, he was arrogant anyway ... Harmon just couldn't get along with those (Redwater) Ukrainians. And those Ukrainians had never before owned an acre of land in their lives. And land was sacred to them! They would not turn loose of it! And anybody stuck anything in the ground, it was like piercing their body. He just riled them up! And, actually, it's a wonder he didn't get shot ... That was dangerous ... Harmon went out there and called all these landowners together, and said 'Now, I'm gonna give you one more chance to sign these documents' and none of them would sign.. And he said, 'Alright ... but you're gonna get screwed anyway' and he told 'em about the Arbitration Board. He said, 'Now, you can't do anything about it. We can go on there - all we have to do is make application'. Well, brother, they got him out of there in a hurry! And that's when Jim Statler said, 'Bill, you've got to go up there and do something about this'...
>
> ...So, I went up there, and the first thing I did was find out who the Member of the Legislature was out of Redwater ... Peter Chaba. I went to him ... explained what I was going to do. He said, 'Thank God we've got somebody that will understand.' He was the nicest fellow ... a friend of mine from there on ... He and I called a meeting ... And he went out, by himself, and talked to them ... If it hadn't been for him...
>
> ...I can remember one woman, Mary Pich! I'll never forget her! She weighed about 200 pounds. Her husband was about 130 pounds. And she was the meanest person. Oh, she'd just try her damnedest to get me riled up, and thank God I never lost my temper ... She was just a bitchy woman! She'd sit across the table from me and she'd say, "Bill Mugler, you're a sonofabitch." ... And she was the sloppiest thing! I remember going up there one day in the summer time. - there were so many flies on that screen door that I couldn't even see in. And that's how her whole house was...

Bill recalls convincing British-American to buy land, then give it back to the people to farm rent-free.

After a brief interlude with a promoter, Bill joined Richfield, with Sam Stewart as manager. Following that firm's linkage with Atlantic Refining, Mugler moved over as vice president, land, for Central Del Rio, with John Hardy as boss. Then word went out that CDR was being taken over by Canadian Pacific Oil and Gas Ltd. When John Hardy retired, Bob Campbell came over, "A very, very good man. I have a lot of respect for him." Bill finally retired in 1974. "Some people wanted me to do some consulting. I said 'no' ... I've worked since I was nine years old at the grocery store on Saturdays."

Winding up the interview,

> ...I don't have much to say. I don't have any excuses for anything that I might have done. I've enjoyed every minute that I spent in the oil business ... If I had it to do over again, I would enjoy doing it - and that's what I said when I retired ... But I want to break my ties, and I did ... I don't think it's near the industry that it used to be, because you can't make a deal on a handshake any more.

JOHN AND OLGA MYROON

"My Dad came to Canada in 1905 from Ukraine to make a better living. He was 19 years of age at the time." This simple, evocative statement by John Myroon conveys the aspirations of these bonded, unlettered peasants. And their hopes were realized in more ways than they could imagine: farm land, a comfortable city home, and above all, "A nice family--grandchildren, great-grandchildren, all doing pretty good."

John's father, Fred, and his bride, Mary (Dereziuk), filed on their homestead, NW5-57-21 W4M. "All the good land was already taken." So Fred had to accept what was left, which was a "sand pile" - marginal land.

John's lean, weatherbeaten face with its infectious smile belies his 80 years and several heart operations. "I was born on the homestead September 7, 1912. I had five brothers and five sisters ... I attended Ufford School and walked three and a half miles, summer and winter. They only taught to Grade IX in the one-room school, so that was all the education I got." John's school teacher came to see his father a year after he quit school, the reason: "No money to send me elsewhere. He told my Dad, 'He's such a bright student'." But this has not proved to be a handicap because, as we shall see, he was a successful farmer and a good organizer among his fellow farmers.

John married Olga (Litwin) June 13, 1940. Olga's parents had come to Canada in 1904. Her father, Stefan, bought four homesteads--one of which was the SE/28-57-20 W4M, her birthplace. "I had seven sisters and six brothers." Olga and John started their new life together on the SW/34-57-21 W4M which they had purchased from Ben Roberts, a disabled WWI veteran. "And the land was so poor, it couldn't raise good crops." Roberts claimed that he could have obtained his mineral rights through the Soldiers Settlement Board, but that unfortunately was an illusion.[13]

John summarizes:

> ...I worked 15 hours every day on the farm ... I have always believed in hard work. It gets you where you want to go ... Friendship is precious, humour is the chemistry of life. I have always had time for fun, for it enriches my life ... Life is meant to be enjoyed always and to the fullest ... Happiness is based on what you do and how you create it, not

---

13 Hubert Somerville checked the chain of title and confirmed that no minerals were conveyed from the Crown to anyone.

> what others give to you ... I look to God for the spiritual part of me and my family...

After acquiring the adjacent NW of 34 the Myroons built their home in 1953. A picture on the wall of their Edmonton house shows the gas plant in the distance, one and one-half miles to the southwest. They purchased their Edmonton house in 1982 and moved into it in 1987.

Olga's philsophy of life hinges on the Golden Rule, "Family is the fabric of society, for it holds us together ... Family is also very precious to me ... The greatest virtue is kindness. If we were considerate and gentle, we would not have poverty and crime." She ends with a plea that we should all seriously consider. "Belief in faith should replace greed and selfishness. Commitment to family and work should be shared by all people."

LAUDER NOWERS

B. L. (Lauder) Nowers was a unique pioneer, born in his father's house in 1910 when that property was out in the country, now in an old built-up section of Elbow Park (3019 - 6 St SW). Having lived in that house all his life, he died in July,1992. Fortunately Harry Simpson and Gibby Gibson had interviewed him in 1987. Here are excerpts:

His first job was in a brokerage house, just in time to be there for the great crash of 1929. We next hear of Lauder in Turner Valley where he had a job roughnecking with Royalite. He boarded at the Black Diamond Hotel and recalled with nostalgia the two sisters who were running the restaurant, Anne and Rose Strilchuk.[14] He would buy a meal ticket for $25, which enabled him to "eat as much as you liked" for one month. Handling 5½-inch drill pipe hardened our hero into shape for WWII, during which he served with distinction.

A charcoal sketch of Lauder Nowers on the occasion of his retirement, 1976.
Furnished by Nowers family.

14 Anne married Al Wright, General Pete's drilling superintendent, and Rose married a Jack MacGregor.

Upon his return to civvy street, "The office (at 606 - 2 St. S.W.) wanted me to go back into the Drilling Department, needing toolpushers, drilling supervisors, even jobs in South America." He turned all this down to work in Personnel. When Leduc broke, Lauder's duties were greatly increased because he had to hire and send men up to George Thompson. With discovery of oil in Redwater in 1948, he recalls being sent out there with "Gibby" (Gibson). "I forget now how many went there at the very first. They slapped together a bunkhouse and an office." In his 1987 interview with Harry Simpson and Gibson, he describes what it was like. "Well, the bunkhouse was really something ... It was colder than the mill tails of hell, and it got colder ... We had no washing facilities. There wasn't any water - red or otherwise. We had water hauled in from Edmonton in big milk cans, and on some of the colder nights, the water in these cans would freeze inside the bunkhouse, not too far from the stove."

When asked how they managed to sleep, "I seem to remember getting into pajamas, but God, it was a struggle. We had about nine blankets apiece, and Gib had Major - his big shepherd dog under the bedclothes that kept him warm. Everybody wore snow boots and, most of the time, an overcoat and mitts in the office ... Pushing a pencil with a pair of mitts was a bit of a trick."

Hugh MacKenzie corroborates Lauder's description.

> ...One thing about the bunkhouses ... they were cold. You never had to worry about getting asphyxiated ... from carbon monoxide from the furnaces, because they never worked ... half the time. But we were in there only a short time, while they were building ... two, big main bunkhouses. And, basically, one was supervisors and management, and one was hourly wage employees. And they had a cooking facility in each bunkhouse so that if you wanted to "batch" you could. Ultimately, they started on an office building for us and two bunkhouses, so eventually we got into pretty decent quarters...

Lauder describes Betty's arrival with the two girls.

> "It was 47 below zero. The Jefferys, next door to us, had slept on the kitchen floor the night before because they couldn't get the house warm enough ... I had an awful time getting some of our furniture in...
>
> Our eldest youngster, Tara, would come home from school smelling of garlic ... These little kids weren't exactly sewn into their clothes, but the whole countryside were poor as church mice when we went in there. Everybody had the funniest-looking old shacks, with cow

> shit mixed with straw as chinking between the logs. Later, the farmers started to collect surface lease rentals, and things improved greatly...

Nowers almost became mayor of Redwater, "But I got transferred to Edmonton just in time." Lauder had much to do with banking, and he remembers the first Royal Bank being managed by Bert Smith. "A hell of a good head. We ... made sure he didn't want for anything. We got him a house." He also describes Walter Malowany, garage proprietor and the Bank's landlord. "He was kind of the head of the Ukrainians that lived around us ... They looked to Walter, and if he agreed, that was it. And if we had a problem, we went to Walter ... You could almost say that it was solved by the time you got it across to Walter what we wanted to do."

THE PASEMCOS

It was a beautiful July day when Jean McLaws and the author visited Victor and Beanie Pasemco in their "home place" on the SE/ 23-57-22 W4M. All the sights and sounds of summer--grand-kids running in and out, the screen door banging, birds singing, and dogs barking. The aroma of baking bread wafted into the large family room. It was a peaceful, homey scene.

Victor and Beanie Pasemco in front of an early dwelling.
Furnished by the Author

Victor's grandparents had settled near Star (six miles east of Bruderheim). He pointed with pride to the small white building just outside his present dwelling, where he was born in 1923. He evoked the instinctive yearning of his kind for the "home quarter" - the indissoluble link, stronger almost than blood, equally shared with Beanie. "...It's a bit of an obsession with me. I don't know what it means to move to or live in another place. I have been blessed with nine grandchildren ... I have literally travelled only a few feet."

Victor left the farm to be Imperial Oil marketing agent, elevator operator, then labourer for Interprovincial Pipe Line on the 50-acre fraction that his Dad sold them. This was followed by a stint at Imperial Oil's gas plant.

One day in 1981, he decided to quit. "Money isn't everything." But the next morning, "There was a feeling of emptiness, because I missed my buddies and the social life in the gas plant." Since then, this void has been filled with a hobby - restoring old furniture and, more recently, travel. Victor is also experimenting with new types of grass for seed purposes, an inevitable return to the soil they instinctively loved.

When asked about his philosophy,"...We take a road that is unpredictable ... some painful experiences ... The reward is, I'm here today. All I want in life is a few more years."

THE ROYER FAMILY

In a March 1992 interview, Yvonne Royer (nee Nadeau) traced her early history. Born in Rhode Island, U.S.A. in 1904, she came out with her family to Val Soucy, Alberta in 1909 to live with her grandparents, who had preceded them. Quite a few Frenchmen were located here: the Odettes, the Paquettes, the Soucys. Indeed, the Soucys had come from Kankakee, Illinois.

Yvonne states, "I didn't have much of schooling ... I went to Woodgrove school, four miles each way, but in winter the roads were so bad that I went to the neighbours twice a week for my schooling." Yvonne recalls going to Edmonton with her father in a big wagon with a team of oxen which would take them all day and more. "...We'd go as far as half-way and stop. Then the next morning, we'd go to the hotel on 101st Street." She describes her father's trade: a blacksmith working for Mr. Desrosier at Lamoureaux.[15]

> ...When I was 17½ I married Emile Royer in 1921. Emile was born in Ft. Saskatchewan in 1900...

Yvonne recalls the Depression days on her Dad's place, bought in 1922.

> ...We were froze out and hailed out ... We couldn't hardly make it ... so we went to the Municipality ... They gave us - the five of us, to keep - $7 a month. That was in 1936 ... It was 1939 that we moved in to Redwater ... Emile started with charging batteries ... And I'm telling you, the food wasn't too much on the table at that time. Then I got in with Edmonton Produce (the Superstein Family). They come out and they wanted us to start up a grading station. Well, I didn't know anything about grading eggs or running a grading station, so I had to go to school to learn the trade...

> ...My husband went out with his old truck, picking up eggs. Then Superstein wanted us to get chickens. So we were buying chickens and turkeys. Then Rita and I packed turkeys in crates at night...

Rita, who was at the interview, recalls the Heiland Crew and the party chief's name, Austin. This was the crew which did the detail work prior to starting Redwater No. 1. Yvonne describes when they struck oil. "It was 1948, and we were in the States on vacation ... It was at that time that Rod Morris rented the chicken house at the back of the property for an Imperial Oil office. Emile built a big chicken

15 Pronounced locally LA-MORE-EE.

house outside the town and eggs were being graded in the house on the street."

Rita remembers Mike Woodhead, who lived in the back room, with his train lay-out. "He was a real model train buff, and even had his own lathe for turning metal parts." Yvonne: "He was tall and, oh, so polite ... There were seven guys that boarded for three months." She recalls her early days, working for Avenue Grocery, marrying Ken Lonsdale in 1949 (one of the sample washers) and raising two children. Her husband worked for Rod Morris in the Geological Department. He later switched to the Production Department. Then Ken was killed in a car accident in 1959. In 1960 Rita went to work for Imperial as a switchboard operator.

Yvonne Royer and daughters, Eve and Rita.
Furnished by Author

Eve Forry (previously Tringale nee Royer) did her own taping at her home in California and sent the results to Calgary, where we transcribed them. She explained how she had been supposed to learn to grade eggs at Supersteins during the War. She met an American who was working on the Alaska Highway "...And that was the end of the egg grading." They moved to the U.S., but the marriage later failed.

Eve remembers Lauder Nowers (Personnel) driving her in to Edmonton to select furniture for the two houses set aside to serve as female "batches". She also recalls Gordon Copland, geologist, who served as cook in the chicken coop.

Mr. Royer died in 1977. His obituary[16] states that he built the chicken house in 1943. The editorial which accompanied the account of his life overstates the chicken coop's functions, but it gives the outsider's perception of what went on behind those closed doors:

> "...The oil company stayed two years in that house, its smallness disguising the momentous decisions made within involving millions of dollars. He (Emile) was at once its history maker, as well as its historian, having lived the longest in Redwater. During those heady months in 1948, it was the chicken house that became the center for a dramatic tale unfolding, of daring detective work in the underground, of a gamble, and in the end of a virtually priceless catch. Unpretentious, as far as chicken houses go, this coop found its way into the local history with a dignity unsung but surely equal to that accorded architecture of greater vintage. This house became the centre for Imperial Oil operations when the company nosed out the edge of an oil pool, laid down money that it was a field, and caught both a windfall and wealth of the imagination of the continent. The house was a place - literally - for chickens, but no place for the chicken-livered, because it became the house for some courageous decisions by Imperial Oil. By the end of 1948, they and other petroleum companies who picked up the scent worked feverishly to prove the gamble right..."

The article goes on to quote the January 8, 1949 edition of the Edmonton Journal as to how Pacific and Anglo-Home achieved success,

> "...Thus, 28 years ago the Edmonton Journal documented the opening chapters of an incredible tale which began its telling in a chicken coop. The historic chicken house has long since been removed, its slumber quite intact on other ground near town and quite unnoticed for all its history. But it had its great day for the record. It was an important nerve center for an oil company that saw the birth of a new oil field in a world yet to feel the crunch of expendable resources."

16 The Redwater News, July 22, 1977.

NICK ROZAK

This colourful personality, born in the Ukraine, spent WWII in a Prisoner-of-War camp working in a creamery and entertaining with his musical prowess. He 'jumped from the frying pan into the fire' when the Russians took over the camp.

Shortly after arriving in Canada Nick got a job with the Edmonton Dairy Pool, "...And the boss told me, 'I would like it if you worked in the Leduc Creamery'." It didn't take long for Nick to learn that roughnecks were getting twice his wage, and he contacted Harry Webster - who had just moved his rig on to the Serenk property for Royalite. "My friend asked Harry, 'How would you like to take this man for a roughneck'?

'Oh, he can't speak English?... Can he operate a steam boiler'?"

One day, Harry said, 'You go work on the rig. I'll get an older man (for the boiler)'." (He continued to roughneck with them until 1957.

Later, after many moves around the province, Katherine, his bride, commented, "Seems you got married with the rig, not with me." "And I said, 'Well, okay, if you don't want it, I'm finished ... I won't go' ... I phoned Harry and he says 'I'll come and talk to you.' But I quit."

Nick and Katherine Rozak proudly displaying his collection of hand-made violins. in their Edmonton. home.

Furnished by Author

Rozak is now retired and spends much of his time crafting his beautiful violins. He recalls forming an orchestra in the prison camp. "And I went with a guard ... about four violins, four violas, and two guitars ... We played pretty nice ... The German guards say, 'They're nice artists'."

The author's interview of Nick ended with "Well, Canada is the best country in the world. and I couldn't make it really good like I made it in Canada ... I told my wife this quite a few times.."

MARY SAWKA

The author had planned to interview Mrs. Mary Sawka for some time, and a meeting was arranged for November 1992. He was greeted with a friendly, open smile. Her eyes say it all! When this shy widow in her late autumn years smiles, it is with a genuine warmth. And yet behind all this is the determined look of a mother who, in the face of adversity, was going to make sure that her children received a good upbringing and the education that went with it.

Mary Sawka at home.
Furnished by Author

Mary Sawka's spotless home, inside and out, is curious in that its main door faces the back yard. Is this because it is on the busy corner of 49th Street and 50th Avenue? "Forty-eight years since I'm in this house." Mary has refurbished this ancient building to modern standards. Built many years ago, and previously occupied by a shoemaker, the china door knobs attest to its vintage.

The conversation was restricted by Mary's hesitant answers. (Were the questions too searching?) She had warned at the outset that she wouldn't be of much help; nevertheless, some interesting facts emerged.

Mary Prokopiw was born in 1911, the daughter of Sophie and John. She had three brothers and a sister. Her parents had homesteaded in the Redwater district. Commenting on her educational level, "Well, I didn't pass Grade IX ... so I didn't go any more." She worked at home until she married Anton Sawka.

"I was married ten years, and he died." A terse statement, tragic in its finality, especially since Mary was left with four young children.

After spending one year in her mother-in-law's house (who was also widowed) Mary moved to Redwater in 1944. Mrs. Sawka recollects little of the beginning of the oil boom. (How would she, being preoccupied with four young ones?) When asked who looked after her children, "I looked (after them) myself - we had no baby-sitter," the reply heavy with meaning, sacrifice, and dedication to her brood.

The discussion then turned to her children: her eldest, Bob, died - leaving behind his widow and his three children: Ray, a successful senior executive with an intermediate oil company; one daughter, a lab technician at University Hospital and the other a secretary at the University of Alberta.

To help out with the chain of the family we referred to a photo on page 183 of "Memories" which shows her late husband, Anton, as a child with the Sawka family. Mary's mother, Sophie Kutney, was 16 when she came to Canada. Born in 1886, she lived to the ripe old age of 99 - passing on in 1985. Mary explained that, until a few years ago, Sophie had lived with her daughter, but her last years were spent in a nursing home.

Mary's living room has many family pictures, and she is justifiably proud of all of her children and their accomplishments. When asked why she had not travelled to the Far East to visit her son, Ray (at one stage a senior executive in Indonesia), "No ... I don't like air travel." Nevertheless, there was no doubt that here was a strong pioneer personality living in the present and future for her offspring.

STETSKO

"...And he loved the oil wells ... the oil work ... that was his life ... he loved it," so mused Helen Stetsko as she recalled her first husband, Alex Kenyon. His short career came to a tragic halt at Devon, Alberta, on February 13, 1951. Over-work had taxed his heart, and having been gassed in WWII hadn't helped.

Quoting the eulogy,[17] "...Alexander Archibald Kenyon was born at Clive, Alberta, in 1918. In the spring of 1937 he went to work at Turner Valley under the able tutelage of his uncle, Sandy Addison. Starting at the bottom in what is known in oil parlance as 'roughneck' ... he received promotion rapidly to the position of Head Driller."

Helen Hlushko, his intended bride, grew up near Wostok and after completing her Grade IX "...Went out working ... too poor ... we couldn't afford to go (any farther in our education). And then I married my neighbour, Alexander Kenyon - Sandy for short - and we moved to Turner Valley for Imperial Oil." Their November 1941 marriage began in Turner Valley, but they soon started to travel when Alex joined the Army in the spring of 1942. Helen recalls moving from town to town, then to Vernon, "Where our first child, Diane, was born." After his discharge, Alex returned to Imperial Oil and roughnecked in the area northeast of Edmonton looking for gas. "He had built a two-room skid shack when we were in Minburn, and we moved it each time a well was drilled."

His old toolpush, Fin Lineham, and Millard van der Grift (and Fin's brother, Jim, as a silent partner) joined up to form Kenyon's Oil Well Servicing Co. Fin recalls, "There was all kinds of work. They were gonna keep Kenyon so busy he wouldn't have time to spin around ... and they did." Then, "The first rig we bought was a Cardwell Mobile Hoist because Kenyon was a cable tool man, and he wanted one with a spudder on it." Before the end of the year, the firm ordered another, smaller one: a Model H rig, which would pull sucker rods and pumps.

Returning to the eulogy "...As general manager, he drove himself to the limit. Unknown to himself, his health broke down ... He suffered a heart attack at a friend's home in Devon. He passed away without regaining consciousness immediately after a stroke."

17 Unknown author.

This was a terrible shock to his young widow, especially since Kenyon died intestate. So Fin Lineham, along with Sandy Addison, had to work through the Public Trustee to cash out Helen's interest and pay her. Lineham remembers, "I agreed to give 25¢ a share more than the price was set at. That's what we agreed on. I got hell for that from Jim and van der Grift ... for spending their money."

Helen has a different story. "I still had my share, but there wasn't any money coming in from the Estate. And they were after me to sell my share. Well, there was no way ... I couldn't keep it because they told me it would be sold anyway. If I was against selling it, they'd still..." Her voice trailed off as she relived those uncertain days with little understanding of the world of business.

The old story of "big is not beautiful" came true with the bankruptcy of the company in 1966, 15 years after Sandy's death. As Fin states, "We were just so far in the hole..." Fin blames the Treasury Branch, "They'd convince them they should enlarge and then, the first little thing that went bad..."

Helen was now left with growing children: Diane and Dennis. She continued to live in Redwater. She describes how she met Matt Stetsko, "We had an oil stove that would freeze up. So I thought, since I was left a widow ... it wasn't going to work ... I thought I better change it to gas ... so that's how the gas man came along." Matt was the Redwater Utilities service man. In due course, they moved to a more comfortable home and raised a new family.

A word about Matt: After mining at Yellowknife, then in the Coal Branch, he arrived in Redwater in 1950 to work for Wilder Ripley's gas utility. The outfit sold the system to the town. Matt describes the earliest days, "A lot of times in the winter, the wells would freeze up. I had to put glycol in them."

Matt boasts of being the oldest Lions Club member in Redwater, 42 years perfect attendance. Busier than ever in retirement - the president of the Pioneer Club and the chief pancake cook. He boasts, too, of his physical prowess, "They can't keep up with me ... I'll be 77 this fall." Helen, in her quiet unassuming way yet with many hidden strengths, stated, "I stayed at home and looked after the children ... helped him with Utilities, answering phones. We always grew a big garden, and I did all the canning. I always worked ... and enjoyed it - enjoyed all my family."

Thus ends the story of a family whose name was still popular in the oil patch for years after Sandy's untimely death in 1951.

APPENDICES

LIST OF APPENDICES "REDWATER"

I	Donors to Re-Erection of Redwater Derrick (L.Chekerda)
II	Redwater "C" Seismic Prospect (Rabey Memo)
III	Wildcats Spudded Before Redwater Discovery
IV	Recommendation, Imperial Redwater No.1 (E.W.Shaw)
V	Redwater "A" Seismic Prospect (Rabey Memo)
VI	Expenditures: Imperial Redwater No.1
VII	Imperial Redwater No.1 Crew - 1948
VIII	Darwin Bibliography
IX	Andrichuk's Appraisal of the Green Hornet Report
X	Baugh's Reservoir Parameters
XI	Fournier's Reservoir Parameters
XII	Haskett: "Redwater," CIM Bulletin, April '51
XIII	Green Hornet Report: Author's Dissertation and Sample Pages
XIV	Gas Plant Start-Up Personnel, 1956
XV	A 1970 Pool History of Redwater (Imperial Oil)
XVI	Deed of Surrender (Hudson's Bay Co. to Dominion Crown)
XVII	Homesteaders on Crown Land, 1894-1905, and plat showing their locations.
XVIII	Redwater Crown Reserve Sales ('48-'52), H. H. Somerville
XIX	Ukrainian Beliefs (and) Yachimec's Brief Historical Review
XX	Book Review, "Memories" 1972

Appendix I

DONORS TO THE REMOVAL AND RE-ERECTION OF THE REDWATER DERRICK

(Larry Chekerda, Ramrod)

Government of Alberta	$5000
Esso	$5000
Redwater Water Disposal Co.	$1000
PetroCanada	$1000
Gulf	$1000
Adam Crane Service (drove the piles)	
Peter & Diane Romanchuk	Cash
Sandy & Barbara Stetsko	
Redwater Machining & Mfg. (drilled holes and cut plates)	
Kenneth and Irene Lloyd	Cash
Charity Construction (hauled the derrick)	
Redwater Lions Club	
Dabco Construction (electrical work)	
DanTom Construction	
Kenting Drilling (crown blocks and generator)	
Sparrow Electric	
Mr. & Mrs. John Chekerda	Cash
Larry Chekerda	Cash
Matt & Helen Stetsko	
SAAN (all proceeds from a weekend sale)	
Walter Ciepliski	Cash
Carriage Real Estate	Cash
Home Hardware	
Tartan Construction (hauling derrick, some welding)	
McKinie Machining, Edmonton (travelling blocks)	
P. Yaworski Holdings (cement, re-bar and cash)	
Acklands Hardware (paint)	
Northwood Auto Glass & Repairs	Cash

Appendix II

REDWATER "C" SEISMIC PROSPECT

(Rabey Memo)

Wes Rabey furnished the author with a copy of his recommendation to drill Redwater "C" anomaly: Imperial Fedorah No.1 in Lsd 13-22-57-23 W4M. Although the report is quite technical, it will prove of interest and value to those earth scientists who wish to know how 1948 geophysics worked.

The "C" prospect was on a kind of pimple. But between the Redwater location and Fedorah there was an increase in the velocity of the Lower Cretaceous shale as it went westward and basinward. This created a spurious anomaly which was reckoned not to be much lower subsea-wise than Redwater. The other reason, according to Rabey, was that Imperial were desperate to acquire the Eric Harvie freehold parcels in 57-21 and 56-21[1] and wanted to delay drilling "A" prospect. Fedorah proved to be on the D-3 reef, which gave Walters more confidence in moving up to drill Redwater to the east.

Rabey's memo stated "Located on the west flank of the main Redwater anomaly on a subsidiary structure ... This test will check the possibility of oil accumulations, as well as give valuable stratigraphic information, which should be a preview of conditions expected when the deep test is drilled at Redwater." There were three reflectors: Second White Specs, Devonian and Elk Point salt.

Rabey then goes into a series of conjectures based on the rather small differences in velocities. He concludes, "...Although the exact position of the Devonian reflection is not known, it could be possible that it represents the attitude of a reef, perhaps the D-3 zone. It is peculiar in that this horizon can be carried over the whole Redwater structure. But to the north and east, where the critical dip occurs on shallower and deeper beds, this reflection disappears - so there must be some lithological change in the section..." He cites, "...Steep east dip on the east flank of the Morinville structure ... however, no evidence of faulting can be detected. Continuous profiling is recommended." Fedorah was spudded on March 1, the centre of a

1 See Chapter 6 for details.

"4x4" crown lease block just selected. There was, thus, adequate land protection even though Imperial took a 4x4 rather than a 3x3 - which they would have had to do after March 29, 1948. The well was completed on June 15, 1948, over 1000 ft into the Elk Point at 6500 ft and was abandoned. Its importance lay in the fact that there was water-bearing D-3 reef present.

Mr. R. A. Walters January 29, 1948.

W. N. Rabey

Report on Redwater "C" Prospect

A test well is proposed to be drilled in Lsd. 13, Sec. 22, Twp. 57, Rge. 23, W.4th Meridian. This is located on the west flank of the main Redwater anomaly, on a subsidiary structure, which shows approximately 0.010 second closure on the deeper horizons and 0.005 second on the shallow ones. This test will check the possibility of oil accumulation in this area, as well as give valuable stratigraphic information which should be a preview of conditions expected when a deep test is drilled on Redwater. This location is approximately 0.050 second (250 to 300 feet) lower structurally than the proposed location on the main Redwater feature, but is still above the steep flank which occurs farther to the west.

There are three reflecting horizons carried on the area:

1. "G" Bed (second white specks approximately).
2. Bed in Devonian (300 to 500 feet below top of Devonian when correlated (?) to Morinville).
3. "N" Bed (top of Devonian salt approximately).

In this area, the Lower Cretaceous reflection shows very poor development and, in most cases, is not present. In estimating the depth to the top of the Lower Cretaceous, there is a possibility that the cycle picked could be a half-cycle low, in which case the estimation would be about 50 feet too deep. However, the interval between the "G" bed and the "R" is the same as at Legal, which would lead one to suspect that the correct cycle has been picked.

The records were quite good, except for a few on the diagonal line in the south-central part of the map area and the occasional record elsewhere due to a shallow hole or muskeg condition in local areas. The time ties on the records were good.

Considering a section west to east across the north end of the prospect, and extending it east to the Redwater structure, the following are found:

From the point H6-1278 (N.W. Cor. Sec. 31, Twp. 57, Rge 23, W.4th Meridian), east 2½ miles to point H6-1258,

Memo: Mr. R. A. Walters. - 2 - January 29, 1948.

The second white specks (approx.) rises ... .019 sec.
Bed in Devonian rises .... .030 sec.
Top of Devonian salt (approx.) rises ...... .037 sec.

This means there is a thinning of 0.011 second (50 feet approx.) in the Lower Cretaceous section, or the upper part of the Devonian, or perhaps part of both, in a west to east direction. There is also a 0.007 second (35 feet approximately) thinning in the Devonian section, in a west to east direction between the above points.

Carrying this section three miles farther east to point H6-1245, on the east side of Sec. 36, Twp. 57, Rge. 23, W.4th Meridian, it is found that,

(1) the section between the second white specks and the bed in the Devonian remains the same.

(2) the section between the bed in the Devonian and the Devonian salt thins by another 0.022 second (100 feet approximately).

When this section is carried another eight miles east to the point H6-929 on the Redwater structure, it is found that,

1. section between second white specks and bed in the Devonian thins by 0.020 second (100 feet approximately).

2. section between bed in Devonian and top of the Devonian salt thickens by 0.012 second (60 feet approximately).

If the correlation is correct and the interval velocities remain constant, the following deductions can be made:

(1) There is a definite thinning in the Lower Cretaceous or the upper part of the Devonian or both, from the N.W. Cor. of Sec. 31, Twp. 57, Rge. 23, to the S.E. Cor. of Sec. 29, Twp. 57, Rge. 22 (14 miles) of about 150 feet.

(2) If the Devonian bed carried is not an eroded surface, the area around point H6-1245 (Sec. 36, Twp. 57, Rge. 23, W.4th Meridian) must have had less deposition in Devonian times, or been subjected to greater compaction as there is an increase in section above the Devonian salt both to the east and west of this point.

(3) If the Devonian bed carried is an eroded sur-

Memo: Mr. R. A. Walters. - 3 - January 29, 1948.

face, then the area around point H6-1245 must have been subjected to greater erosion than were the areas to the east and west.

Considering a section northwest to southeast, from the point H6-1267 on the north side of Sec. 29, Twp. 57, Rge. 23, W.4th Meridian, south two miles to the point H6-794,

(1) The reflection representing the second white specks (approx.) rises .... 0.010 sec.

(2) The reflection representing a bed in the Devonian rises .... 0.027 sec.

(3) The reflection representing the top of the salt (approx.) rises .... 0.039 sec.

This means:

(a) There is a thinning between the second white specks and the bed in the Devonian of 0.017 second (85 feet).

(b) There is a thinning between the bed in the Devonian and top of Devonian salt of 0.012 second (60 feet).

Following this still further, a definite thinning of approximately 0.020 second (100 feet) can be seen in the upper part of the Devonian section. This does not appear to be due to frequency change, as groups of reflections both above and below this Devonian group maintain their time interval.

Although the exact position of the Devonian reflection is not known, it could be possible that it represents the attitude of a reef, perhaps the D_3 zone. It is peculiar, in that this horizon can be carried over the whole Redwater structure, but to the north and east where the critical dip occurs on shallower and deeper beds, this reflection disappears as though there must be some lithologic change in the section. This could be caused by some limestone or dolomite grading into shale where the section above and below is shale. However, to the south and west this horizon carried down the steep flank as far as we have carried our continuous profile. Why this reflection should carry down these flanks to the south and west and not to the north and east is difficult to explain, unless deposition or reef development happened before the sinking to the south and west occurred. There might be such a condition as a graben or keystone fault, because there is steep east dip on the east flank of the Morinville structure. However, no evidence of faulting can be detected from present work. It would be desirable to have continuous profiling to determine if the reflections from the Redwater structure carry through the syncline or graben and tie in to the reflections on the Morinville structure.

W. N. Rabey

WNR/dic W. N. Rabey.

Appendix III

WILDCATS SPUDDED BEFORE REDWATER DISCOVERY

		Spud Date	FTD
Clonmel Bruderheim	4-32-55-21-W4	Apr.__, 1942	3212
Anglo-Home Fort Augustus	7-29-55-21-W4	Oct.__, 1946	3313
Amerada Crown	13-21-57-19-W4	Aug. 30, 1948	2825
Imperial Oil wells:			
Morinville No.1	9-30-56-24-W4	May 18, 1947	5078
Legal No.1	8-18-58-24-W4	Sept 3, 1947	6500
Legal No.2	1-17-58-24-W4	Jan. 5, 1948	4008
Fedorah No.1	13-22-57-23-W4	Mar 1, 1948	6500
Volmer No.1	2-16-55-25-W4	Mar 11, 1948	7272
Ardrossan No.1	8-17-53-21-W4	July 18, 1948	7801
Morinville No.2	8-11-56-25-W4	Aug 25, 1948	4320
Mearns No.1	15-06-56-26-W4	Aug 30, 1948	8314

WILDCATS DRILLED PRIOR TO REDWATER DISCOVERY

What was the main thrust behind these wildcats, all of which turned out to be dry holes? Ray Sluzar, after Leduc, said geophysics ruled the roost. Not that the anomalies were any better, many of the wells had been drilled on wide contours or on one-shotpoint highs. One must hark back to O. B. Hopkins' order that all wildcats had to be located on seismic features. After all, what else was there? Correlation that could have been obtained from the core hole drilling of 1945 and 1946 was available, but to what extent was it utilized?

All of the drilling recommendations that the author was able to put his hands on (some of them very sketchy) were cautious and not at all optimistic. Indeed, the geophysical anomalies were usually within the limit of error in terms of velocity contrast. In other words, anything with less than 50 or 60 ft. of relief could be easily ruled out as being spurious.

For instance, the recommendation for Imperial Volmer No.1 in 2-16-55-25, spudded March 11, 1948, was not a very hearty one, there being only 40 ft of indicated closure on the seismic map. The prognosis referred to water in Morinville No.1, ten miles up regional dip and, as was to be expected the D-3 yielded 3000 feet of water in 30 minutes on DST. The hole bottomed in Cambrian beds at 7272 ft. Attempts were made to complete the Ostracod Zone as an oil well, but failed. Why was this well drilled? To validate new land or to keep permits/ leases in good standing?

A somewhat longer recommendation written by Barney Clare (signed by J. B. Webb) was not much more encouraging for Imperial Ardrossan in 8-17-53-21-W4M. The geophysics must have shown up something more interesting because Frank Roberts' reconnaissance was followed up by Heiland detailed shooting. "It is mainly from the seismic work that the location is recommended," but the 75-ft closure is almost within the limits of error. Little hope was held out for D1/D2/D3 at Ardrossan. "A large amount of money has been spent in exploration but the location is lying in the speculative category." The results bore out this forecast; it was a self-fulfilling prophecy, and the hole was abandoned.

The land regulations had not started to be toughened up until later on in 1947 and early 1948, but maybe Imperial Oil felt the need to evaluate these hickeys to see if there was another Leduc out there - to be sure that the company was meeting its drilling obligations, which had not as yet been formalized, and to validate acreage.

And even after Redwater was discovered the number of dry holes continued in that area to the west of Redwater over on to the Morinville reef. There was one benefit: lots of new stratigraphic information which would keep the backroom boys busy for quite a while.

Outside of Clonmel Bruderheim, drilled in 1942, the closest dry hole to Redwater was Anglo-Home Fort Augustus 7-29-55-21, spudded October 9, 1946, possibly without benefit of seismic.

IMPERIAL LEGAL No.1 - 8-18-58-24-W4M

Commenced drilling September 3, 1947. Tested gas on the way down but the D-3 was water bearing. Unfortunately, the Imperial Oil well files are devoid of any definitive data.

IMPERIAL LEGAL No.2 - 1-17-58-24-W4M

One mile to the east, it was commenced immediately upon completion of No,1 using the Franks #3 Rig. The reasons for drilling were given: "Commercial gas in the Viking and lower oil shows at No.1. Seismic maps suggest a higher position on the Legal structure, the magnitude of which increases with depth". Results were no better in the D3, almost flowing water on DST. Despite the good gas flows in the Viking the well was abandoned and most of the long string pulled out.

IMPERIAL MORINVILLE No.1 - 9-30-56-24-W4

The first well to be drilled following the Leduc discovery. Spudded May 18, 1947. It was abandoned in the Devonian at 5078 ft. As with Legal #1, there is no record of this wildcat in the Imperial Oil files.

IMPERIAL MORINVILLE No.2 - 8-11-56-25-W4

Four miles southwest of #1, #2 was drilled over a year later in August, 1948. The recommendation is contradictory in that a "seismic anomaly is present but there is no sub-surface control. It is interesting that this wildcat was drilled and abandoned at 4320 ft using the Franks #2 Rig, Harry Webster, Toolpush. Water in the D3 and gas in the Viking.

IMPERIAL MEARNS No.1 - 15-06-56-26-W4

About eight miles west of Morinville. Was spudded the very day Redwater made its first oil DST. It took nearly four months to drill to 8316 ft into the Precambrian, a 4" core confirming banded gneiss. The cost of drilling was estimated at $190,000 with the usual vague comment "seismic anomaly" and "water only in Devonian rocks from wells in the same general area". Why drill a deep test? Was it Ray Walters' insistence on obtaining deep geophysical information, or was it the last kick at the cat by "Doc" Link before he left the company?

IMPERIAL FEDORAH No.1 - 13-22-57-23-W4

About ten miles west of Redwater village, was spudded on March 12, 1948 and abandoned June, 1948 at 6500 ft. after having tested small amounts of water in the D3.

Fedorah[1] was to be drilled at almost the same time as Volmer using Franks #2 Rig (Blinn pushed operations out of Legal, ten miles to the West.

"Detailed seismic coverage" - "seismic structure" (undifferentiated) was the only "given" reason. Located on a Crown lease block (16 sections). The last objective was to be the "salt" and the porous zone below that.

AMERADA CROWN - 01-21-57-19-W4M

Company formed in 1920 (America-Canada) by DeGolyer.

Ken Reed and A.T. Jacobson were the company senior officers with offices in the Blow Block, 212 - 7 Ave SW (across from the Bay), Calgary. They had set up shop shortly after Leduc but had not participated (as far as we know) in Crown Reserve Sales.

They did have a spread of Crown lands east of Redwater and drilled seven wildcats commencing in June 1948 and completed in December. The closest of these was "1-21-57-19" in lsd 13-21-57-19-W4M spudded August 3, and abandoned August 19, 1948 at 2875 ft. No markers are available but it is believed they were close to the D3.

The company had been negotiating with Harvie as to his mineral titles (see Chapter 6) but were unsuccessful. They would score in June 1949 when they purchased Sec. 16-57-21 for over $3 MM from the Crown.

---

1 IMPERIAL FEDORAH #1 (Origin of Fedorah - "H" presumably added later). Victorien Sardou, French playwright, wrote "Fedora" in which Sarah Bernhardt performed. The year 1908, when the name was used, was the same year Sardou died at the height of his career.

CLONMEL BRUDERHEIM - 4-32-55-20 W4M

About four miles south and east of what would become the southeasterly limits of the Redwater Reef. Clonmel was an Ontario corporation (1942) with offices on Bay Street {The Financial Post Survey of Oils shows it to have held 2,500 acres of land, (Crown)}. The directors were all from Toronto except for Arthur Fenwick Mewburn, listed as Field Manager, with headquarters in Calgary. It was Art's idea to drill this well. There is no indication as to reasons for location. Mewburn was listed on the application for licence as being the Managing Director, with two addresses: Macdonald Hotel, Edmonton, and 436 Lougheed Bldg., Calgary (the offices of Porter, Allen and Millard, now MacKimmie Matthews). The drilling licence was approved by J. J. Frawley, then Deputy Chairman, Petroleum & Natural Gas Conservation Board (now ERCB) December 31, 1941. The drilling contractor was General Petroleums with a very primitive set-up, one of the drillers being Al Wright. The well was spudded on January, but did not actually get going until April with a change of rig. Progress was sporadic and finally gave up the ghost in September at a total depth of 3212 ft. "This well has been abandoned after plugging back to 2976 where a negative drillstem test was made. There were also a series of perforations".

The Abandonment Notice was signed by Mewburn and once again approved by Frawley, now Chairman.

An interesting sidelight is that a Declaration of Expenditures was submitted by J. U. Montalban to enable the company to qualify for credits for lease rentals. The total outlay was $36,400.00 - (about $10.00 a ft.). Of much more interest to the Geologist was the descriptions of the cores showing Dolomite at 3181 to TD, streaks of porosity. Several years later J. C. Sproule of Imperial Oil examined the cores and described the Dolomite as "buff resinous, both crystaline and dense, some good porosity and well developed". Despite the fact that over a thousand feet of hole was cored, the recovery was only 45%. Doug Layer is believed to have made some theoretical comments about the existence of reefal limestone to support his idea of moving west of that dry hole to Leduc. This idea is lost in antiquity.

It is interesting to note that in 1984 Wellore moved into the area and completed a Cretaceous oil well on the same LSD.

ANGLO HOME FORT AUGUSTUS - 7-29-55-21-W4

Was one of the very few wells drilled prior to the Leduc discovery. This was on jointly held lands on which C&E Corp. participated and was associated with the search for gas from the west. It started October 1946 and completed the same month at 3313 ft. It was 12 miles due south of the village of Redwater.

Anglo Home C&E was a triumvirate of oil companies "master-minded" by the late Ray Milner and Major James R. Lowery. This group had drilled a number of wells farther east and had found some Cretaceous gas. There appeared to be no incentive or guide to making the location they did. There was no need to justify locations such as Imperial had to do with seismic anomalies, but that does not mean that the area had not been shot by the group.

Gordon Darling described the Anglo Home cores as being "silty dolomite, good porosity, intergranular". None of this was drillstem tested over the 35-ft. interval. What was it: D-2 or D-3?

There was the mistaken belief that these dry holes were adequately evaluating the area. Nothing could be further from the truth.

Appendix IV

RECOMMENDATION - REDWATER NO. 1

(E. W. Shaw)

LOCATION RECOMMENDATION

IMPERIAL REDWATER NO. 1

Location: Lsd. 1, Sec. 32, Twp. 57, Rge. 21, W. 4th M., Alberta.

Elevation: 2004 feet above sea level.

It is recommended that a stratigraphical test be drilled at the above location to a depth necessary to reach Precambrian rock. The location is approximately 1½ miles north of the village of Redwater on the banks of the Redwater river, a small tributary of the North Saskatchewan river.

In relation to other important stratigraphical tests drilled by Imperial Oil Company, the location is as follows:

(1) 10½ miles east of Imperial Fedorah No. 1.

(2) 20 miles east of Imperial Legal No. 1.

(3) 23 miles northeast-east of Imperial Morinville No. 1.

(4) 50 miles north-northeast of the Leduc oil field.

EXPLORATION

Surface geological studies were conducted along the banks of the North Saskatchewan river in the early part of the summer of 1946 by Mr. J. Y. Smith for Imperial Oil Limited. These studies served to clarify the detailed stratigraphy of the Belly River formation in this general area and assisted in the subsequent interpretation of core drilling data. During the latter part of 1946, the Redwater area was included in a regional program of core drilling. The core drilling program served to determine structural geology in a general way and drew attention to the possibilities of local anomalous structural conditions.

- 2 -

Reflection seismograph surveys were made of the area during the entire year of 1947 and continued to some extent in the early part of 1948. The area was first covered by reconnaissance shooting, and was later detailed with continuous profiling as a result of indicated structural anomalies from the earlier reconnaissance work. In all, a most intensive seismic survey has been completed.

(1) Structural Geology

On the basis of very extensive geological and geophysical surveys conducted by Imperial Oil Limited from 1946 through to the present, a northwest elongated anticlinal structure has been outlined. This structure has an area of approximately 125 square miles and a closure in the order of 150 feet. The regional southwesterly dip is interrupted along the northeast side by a rather sharp reversal. Good seismic reflections were obtained from four stratigraphical horizons interpreted as follows:

(1) Second White Speckled Shale Marker in the Colorado group of Upper Cretaceous age.

(2) Top of the Mannville formation of Lower Cretaceous age.

(3) Approximately top of Jefferson formation of Upper Devonian age.

(4) Approximate top of Middle Devonian evaporite series The structure on these various horizons is considerably different in detail but in good general agreement.

(2) Stratigraphic Geology

The area is immediately underlain by non-marine beds of the Belly River formation. No typical Bearpaw shale is present immediately under-

- 3 -

lying the Belly River formation. A summary of the expected geological section to be penetrated in the drilling of this well is as follows:

Age of Formation	Depth to	General Description
Pleistocene	0'	Boulder clay.
Cretaceous		
Belly River Formation	100'	Fresh water sands, silts, shale and coal.
Lea Park Formation	600'	Dark grey marine shale.
Colorado group	1025'	Dark grey marine shale.
Viking Member	2045'	Interbedded marine sand and shale.
Mannville Formation	2210'	Predominately fresh water sands, silt, shale and coal, with some marine shale.
Devonian		
D1	2965'	Limestone, dolomite and anhydrite.
D2	3035'	Dolomite and silt.
D3	3235'	If present, upper part probably coralline limestone and lower part dark brown dolomite.
Middle Devonian	5235'	Salt, silt, argillaceous limestone, and calcareous dolomite.
Cambrian	6075'	Glauconitic shales, sandstones and limestone.
Precambrian	7500'	Granite and granitic gneisses.

OIL AND GAS POSSIBILITIES

The oil and gas possibilities underlying the proposed location may be summarized as follows:

(1) Viking Sand Zone

Natural gas is known to occur in this zone in the Viking-Kinsella gas field, the Provost gas field, the Leduc oil field, the Morinville and Legal areas, and various other places throughout East-Central Alberta. Additionally, the prospects for obtaining commercial oil production from the Viking Sand are considered reasonably good.

(2) Lower Cretaceous Sand

Thick sand members have been encountered near the top of the Mannville formation in neighbourhood wells. Also a basal quartz sand, probably equivalent to the McMurray tar sands in age, are anticipated. These Lower Cretaceous sands have been non-productive in areas farther south, but with the northerly increase in marine sediments of Lower Cretaceous age, it is hoped that these sands represent distinct prospects at this location.

(3) Upper Devonian Dolomites

It is believed that porous dolomites equivalent in age to the D1, D2, and D3 zones of the Leduc field will be encountered at this location, and that one or all of these zones may be productive of oil and gas. The location, however, is considerably removed from other deep tests. Therefore, satisfactory porosity developments in these Upper Devonian zones is a highly speculative matter.

Porous dolomites have been encountered in some of the places in the Middle Devonian evaporitic series. Very little is know, however, about the stratigraphy of the Middle Devonian and therefore it becomes necessary

-5-

to penetrate this series to widely separated points in order to determine the broad outlines of facies delineation.

Practically nothing is known of the strata of Cambrian age in East-Central Alberta. The prospects for commercial occurrences of oil and gas are not considered to be good in Cambrian strata, but nevertheless drilling through these strata is recommended in order that the entire sedimentary section in this area can be intelligently evaluated.

The drilling of this well, if unsuccessful in terms of finding commercial oil and gas sources, would at least provide a wealth of geological information which should be of considerable value in evaluating other areas in Central Alberta. A large amount of money has been spent in the exploration of this area, and in view of the good indicated structural conditions, subsequent wells in the immediate vicinity may prove to be justified.

E. W. Shaw

E. W. Shaw
P. Engr. (Alberta)

EWS/dk

Appendix V

REDWATER "A" SEISMIC PROSPECT

(Rabey Memo)

Foreword

This appendix is not for everyone. The recommendation to drill Redwater "A" dated July 27, 1948 had the benefit of more detailed shooting (two Heiland crews) than had Fedorah. Weathering (LVL) along the Redwater River Valley gave false readings. As with Appendix II this recommendation presents 1948 technology to 1993 readers who want to know more about the doodlebugs of those days.

It is important to explain to the lay reader that 1000 ft of shale at 13,000 ft/sec has a slower velocity than 1000 ft of reef at 18,000 ft/sec. In 1946 Rabey shot eastward across the Redwater reef, where both the Cretaceous and the Elk Point salt were excellent reflectors. Then, suddenly, the Elk Point appeared to drop. This was due to the energy travelling through first, the fast reef, then at the edge, into slow shale.

Wes credits Bud Coote with improving instrumentation. Old automatic volume control had a delay in its operation so that the blast created by the Cretaceous reflector masked the possible D-3 and, of course, the salt.

Quoting from Rabey's report, "...Located on the top of the main Redwater structure which shows approximately 0.045 seconds closure on the deeper horizon ... this location is structurally higher than either Clyde or Fedorah." Once again Wes cites the three reflectors. Then we mentioned that the records vary from poor to good - ranging from good to the south and west to poor to the north and east. Confirming Bud Coote's comments, two Heiland parties had been put to work. There was a great deal of continuous profiling. Wes goes into technical detail, referring to "Exceptionally good reflection on the southwest side of the Redwater River but energy being poor and subject to "split" cycles and character changes on the north and east side of that same river ... Some change in lithology can be expected." He comments on other reflectors but states that the Devonian salt is the most consistent, maintaining its character extremely well. There is still some evidence that they were thinking "fault". He also states what Chapman probably pointed out - that there may be reflection coming from a steep face, such as the edge of a reef. And he correctly predicts that it would have to "present a plane surface such that waves could be reflected from it." There was some concern that there be D-1 carbonate at Redwater. "The section thickens very rapidly to the northeast, causing a difference of 0.040 seconds from the crest of the structure to the center of the syncline." This was merely the off-reef. This shows an almost complete absence of communication between the Shaw Report and the Rabey write-up, although Rabey does cite geological phenomena to account for the anomaly.

PRODUCING DEPARTMENT
WESTERN DIVISION

MEMORANDUM

Technical Geophysical

To: Mr. R. A. Walters Date: July 27, 1948.

From: W. N. Rabey

REPORT ON REDWATER PROSPECT

A deep test is proposed to be drilled in Lsd. 1, Sec. 32, Twp. 57, Rge. 21, W. 4th M. This is located on the top of the main Redwater structure, which shows approximately 0.020 second closure on the shallow horizon and 0.045 second closure on the deeper horizon. This location is higher structurally than either Clyde or Fedorah wells.

There are three reflecting horizons on the area:

1. Horizon in Lower Cretaceous (approximately 250 feet below the top of the L. C.).
2. Horizon in Devonian - (One cycle below the D3 zone).
3. "N" horizon (Top of Devonian salt approximately).

In addition to these three horizon maps, two isopach maps were made - one showed the interval between the Lower Cretaceous and the Devonian reflections, the other showed the interval between the Lower Cretaceous and the Devonian salt reflections.

The records varied from poor to good. In the south and west parts, the records were quite good but became poorer in quality to the north and east.

Two Heiland parties were used to work out the area and in general their work tied in quite closely. Only on one line H8-541 to H8-547, which was a rework of line H8-818 to H8-825, was there any appreciable difference. This will be rechecked at a future date when this town-

Form 1322 (5m-7-48).

- 2 -

ship will be completed.

The following points are abnormally low on all maps and suggest an erratic L. V. L. condition (e.g. an old stream bed, a fault plane, or some similar condition which allows these points to line up along a fairly straight line). One line extends from H6-977 southwest through H8-122, H8-146, H8-103, H8-73, to H6-883. Another line extends southeast from H8-542 through H8-331, H8-330, H8-98, H6-921, H6-125 to H8-167. These points vary from 0.005 sec. to 0.020 sec. from the average of the points on either side of them. Most of these points have been considered erratics and have been ignored in the contouring.

The Lower Cretaceous reflection represents a horizon approximately 250 feet below the top of the L. C. Where the records are fair to good, it is a very well developed cycle and maintains its character over the whole area. The map indicates the structure as a broad irregular anticline, with approximately 0.020^{+} sec. closure. There is a very prominent low area which tends to split the anticline into two parts along the strike. The exact significance of this is difficult to explain but it might be indicative of a possible fault zone through this area.

The Devonian reflection is one cycle below the D3 cycle, as correlated at the Fedorah well. It is an exceptionally good reflection on the southwest side of the Redwater River but the energy is poorer and it is subject to "split" cycles and character changes on the north and east side of the same river. The amount of energy return from this horizon is probably due to the A. V. C. action of the instruments and the presence or absence of a strong L. C. reflection. However, the character changes and "split" cycles are believed to be due to lithologic changes.

- 3 -

When this reflection group is compared to a similar reflection group at Leduc, it has been observed that the cycles are clean cut with no splits when on the D3 reef, which compares with the area between Fedorah and the top of the Redwater structure. As the edge of the Leduc reef is passed, the cycles are not so well developed with a tendency to split, as is the case on the northeast edge of the Redwater structure. This is not to say that the reef, if it exists, will cease on the northeast edge of the structure but to compare the conditions with those at Leduc and to point out that some change in lithology can be expected.

Although this horizon has been mapped as one consistent reflection, it is not necessary that such is the case. For a considerable time, it was thought this reflection ceased on the north east side of the structure and this could easily be with the character change so predominant along this area. It is possible that the northeast portion of the map is an equivalent lithologic horizon, or even an entirely new one. The lack of steep dip for a reef horizon on the critical side of the structure leads the writer to suspect that this one is not continuous.

Another interesting feature of the Devonian section on the northeast edge of the structure, is the presence of a deeper Devonian reflection, which comes between the reef and Devonian salt. This strong reflection begins on the critical side of the feature and continues to the northeast. The exact significance of this is not known, but it might be a bed butting up against the reef, which was deposited after the reef began growing or even after the reef had ceased. It is strange that as the shallow reflection becomes poor, that this reflection becomes so prominent.

The "N" reflection which correlates with the Devonian

- 4 -

salt at Fedorah, is the most consistent over the whole area. It maintains its character extremely well and the energy return from the horizon is very good over most of the area, except where the deeper Devonian reflection (mentioned above) comes in and masks it. This is considered to be an instrumental difficulty.

This horizon shows more closure than the other ones. It indicates 0.045 sec. to 0.050 sec. with steep dip recorded along the northeast portion of the structure. It defines the structure as a well developed anticline, extending from the southeast corner to the northeast corner of Twp. 57, Rge. 21, W. 4 and continues diagonally northwest to a point half way across Twp. 58, Rge. 22, W. 4th, as it is outlined now.

A second interpretation of the "N" horizon was made, including a fault zone possibility, extending from northwest to southeast on the northeast side of the structure. There was no direct evidence of faulting along this zone. It was postulated from a "disturbance", which crossed the record, sometimes blanking the shallow Devonian and salt reflections. At times this "disturbance" followed the same path as "ground roll" and could be such, but on some records it took a different path. It is possible that the disturbing factor was caused by a L. V. L. condition but it is peculiar, in that it is restricted to such a narrow zone because on either side of this area there is little or no ground roll.

Another possibility, other than being a fault zone, is that there may be a reflection coming from a steep face such as the edge of a reef. The reef, however, could not have many protrusions and would have to present a plane surface such that waves could be reflected from it.

This (fault zone?) only seems to affect the Devonian

\- 5 -

beds, although on the Lower Cretaceous map, a very pronounced low area occurs where the fault zone exists on the deeper beds. The disturbance on the record, however, seems to affect all the reflections, depending on where it begins.

The isopach map, between the Lower Cretaceous reflection and the salt reflection, shows a very uniform time interval, with slight variations up to 0.010 sec. from the Fedorah well across to the northeast side of the Redwater structure. This means if the Lower Cretaceous section remains the same as at Fedorah, there should be some D1 at Redwater. However, the section thickens very rapidly to the northeast, causing a difference of 0.040 sec. from the crest of the structure to the centre of the syncline, which means an increase of between 200 and 300 feet of section.

The isopach map between the Lower Cretaceous reflection, and the shallow Devonian reflection, indicates a thin area over the anticlinal structure, with a slight thickening to the northeast. The thinning from the Fedorah well to the Redwater structure is approximately 0.015 sec. or about 100 feet. This thinning can take place in either the Lower Cretaceous or the Devonian sections or part of both. It indicates there should still be some D1 left at Redwater, providing the Lower Cretaceous section remains the same as at Fedorah.

W. N. Rabey.

Appendix VI

EXPENDITURES IMPERIAL REDWATER NO.1

Form 212A

DOMINION OF CANADA
PROVINCE OF
ALBERTA

IN THE MATTER OF an application made by Imperial Oil Limited of the City of Calgary, to the Government of the Province of Alberta, Department of Lands and Mines, to have certain expenditures made in drilling operations on L.S.D. 1 Sec. 32 Twp. 57, Rge. 21, W. 4 Meridian, being on Petroleum and Natural Gas Lease No. ...., accepted in lieu of rentals by the Government.

TO WIT:

I, Kenneth Preston Jeffrey of the City of Calgary Province of Alberta DO SOLEMNLY DECLARE that during the period from June 1st 1948 to March 31, 1949, expenditures as shown hereunder were made on Well Imperial Redwater #1; That these expenditures have been liquidated and that there is no outstanding indebtedness in connection therewith.

Labour	$ 22,478.27
Subsistence	$ 3,818.70
Employee's Welfare	$ 5,342.70
District Expense	$ 1,745.50
Materials and Supplies	$ 19,945.03
Special Services	$ 26,439.89
Mud Expense	$ 3,996.98
Drilling Tools Expense	$ 18,507.20
Transportation	$ 11,439.85
Geological	$ 7,532.78
Abandonment Expense	$ 1,684.53
Lease and Rental Expense	$ 825.00
Fuel and Power	$ 4,060.00
Miscellaneous	$ 888.38
Camp Expense	$ 1,781.25
Contract Payments	$ 1,847.35
Overhead and Administration	$ 6,271.83
TOTAL	$ 138,605.24

And further, that the following operations were carried out at the well during this period:

AND I HAVE TAKEN THIS SOLEMN DECLARATION conscientiously believing it to be true, and knowing that it is of the same force and effect as if made under oath and by virtue of THE CANADA EVIDENCE ACT.

DECLARED BEFORE ME
At the City of Calgary
Province of Alberta
this 19 day of May
A.D. 1950

K. P. Jeffrey

(A Commissioner of Oaths in and for the Province of Alberta.)

Note:—Use separate sheet for any explanation of items shown above.

Appendix VII

IMPERIAL REDWATER NO.1 CREW - 1948

Drillers
Wilbur Anderson
Earl Bennett
Doug Gamble
Alex Kenyon

Derrickmen
Wilf Bowe
Wyne Green
Slim Owre
Harvey Smith

Cathead
Al Desnoyer
Glenn Gamble
Lou Gattinger
Jim Marcinko

Floormen
Gordon Astle
Norman Gowing
Harold Kinnear
Don MacNeill
Bill Mitchel
Bernie Poliwszyuk
Cliff Rattray
Joe Schlosser
Jack Wenger

Mechanics
Herb Adams
Av Gardiner
Walter Hansen
Gordon MacLeod

Welder
Elmer Gavel

Truck Driver
Wilf Desrosier

Geologist
Rod Morris

Sample Catchers
Tom Charlton
Ab Lonsdale
Ken Lonsdale

Engineers
Jack McCaskill
John Staples
Don Thompson

Mud Men
Don Hunt
Jim Tod

Tool Pusher
Fin Lineham

Amended by Gordon Astle December 1, 1992

Appendix VIII

DARWIN BIBLIOGRAPHY

Darwin, Charles, 1837, On certain areas of elevation and subsidence in the Pacific and Indian Oceans, as deduced from the study of coral formations: Geol. Soc. London Proc, V2, pp 552-554.

Darwin, Charles, 1842, Structural distribution of coral reefs: reprinted 1962 by U. of California Press from 1851 ed. with foreword by H. W. Menard, 214 p.

Darwin, Charles, 1883, On the structure and distribution of coral reefs, also geological observations on the volcanic islands and parts of South America visited during the voyage of HMS. Beagle: London-New York-Melbourne, Ward, Loch and Co., 549 p.

Darwin, Charles, 1896, The structure and distribution of coral reefs, 3rd ed: New York, D. Appleton and Co., 344 p.

Darwin, Francis, ed. 1888. The life and letters of Charles Darwin, including an autobiographical chapter; Lond, John Murray, 3 vols, VI 395 p, V2 393 p, V3 418 p.

Appendix IX

ANDRICHUK'S LONGHAND ANALYSIS

OF THE

GREEN HORNET REPORT

APPENDIX F-1

STRUCTURAL & POROSITY CHARACTERISTICS OF THE REDWATER FIELD

Redwater Working Committee April 15, 1953

Geology

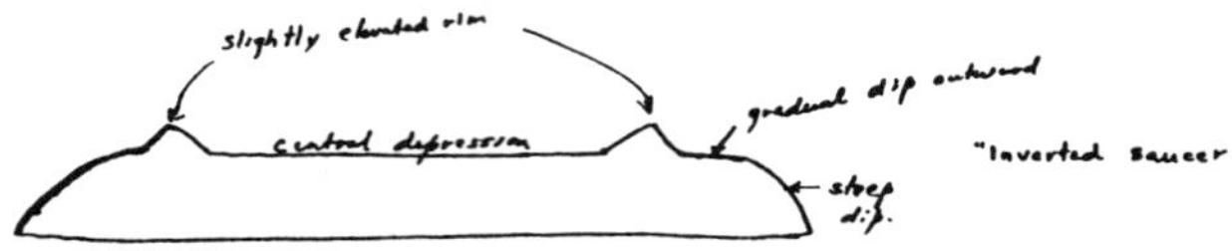

Vertical Cross-Section of Redwater Reef at time of cessation of reef-building

Reef tended to be more porous - from rim → outwards, than underneath central depression

With subsequent upward tilting — of reef to NE — oil would migrate to the uptilted rim, with most of migration taking place along the line of least resistance around the outer edge of the reef. Water drive should be most effective around this porous outer edge.

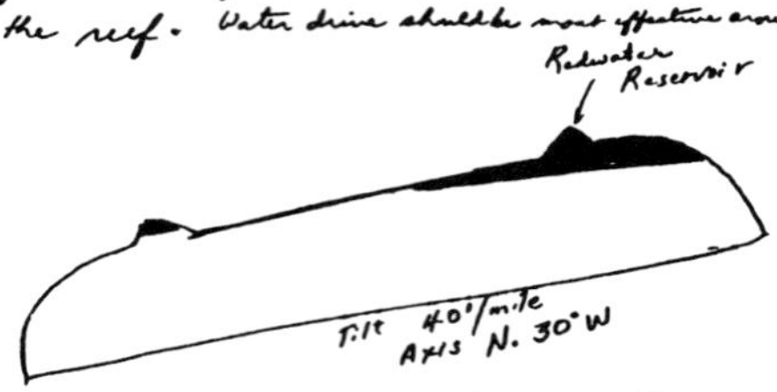

Vertical Section thru Redwater Reef — TILTED

Dynamics of Reef Growth

Wave-resistant structures - formed by processes of

(i) Coralline + other organic frame-building
(ii) Sedimentary retention
(iii) Binding

Organic frame-building occurs prolifically at seaward sides of reef where wave action + currents provide plentiful supply of food to organisms.

Concurrent with reef building — considerable breaking down of structures by wave-action, particularly at times of major storms

Building of reef structures — involved several different environments — rock types in each (facies); certain "facies distribution" pattern. several distinct facies groups → "reef-complex"

F.2

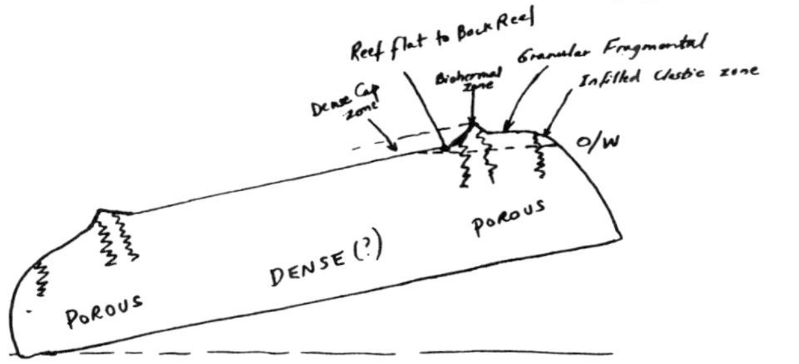

FACIES ZONES

BIOHERMAL ZONE

- zone of major organic growth – facing on the sea
- acted as resistant sea-wall to the force of the waves
- despite destruction it has grown generally upward as the sea rose
- rocks = heterogeneous agglomeration of organic material with abdt. med. to fine porosity
  - organic debris in place caused by reef destruction
  - inwashed granular material
  - dense limestones.

GRANULAR-FRAGMENTAL ZONE

- accumulations of fragments broken from the organic reef. – on seaward side of bioh...
- deposits occurred under turbulent water conditions
- wave-action instrumental in rounding the smaller fragments + grain
- fine lime flushed, so that primary interfragm. + intergranular porosity has been preserved.

INFILLED CLASTIC ZONE

- seaward from biohermal zone, sed. or "clastic" materials change from coarse to fine as depth below sea-level increases.
- when depth greater than that affected by ~~[illegible]~~ wave action, quiet water conditions prevail
- in quiet water – deposition of:
  - finer granular-fragmental accumulations
  - pptation of fine lime, simultaneously.
  - interstitial spaces to be infilled + porosity destroyed

REEF FLAT - TO BACK-REEF ZONE

- behind biohermal sea wall - environment of rel. quiet water -
- encouraged very considerable physico-chemical deposition of fine lime
- considerable to abdt organic growth + scattered shell fossils
- fossils are smaller + frailer varieties - do not contribute as adequately to porosity
- micro-fragm. material - forms a large prop. of rock - caused by wave action over reef wall or thru surge channels
- facies group = variable - but tends to be bedded or "biostromal"
- largely potentially porous, but strong tendency to infilling by fine lime.
- during major storm periods - coarse to fine clastics washed over the reef - causing sporadic deposition on back-reef side.
- contains quickly changing rock types both vertically + horizontally parallel to the reef trend.

DENSE CAP ZONE

- occurred after cessation of reef - building
- predom. fine lime, frequently argillaceous
- differs from most of dense material of underlying zone
- relatively non-organic - potentially non-porous.

POST - DEPOSITIONAL EFFECTS

(1.) Regional tilting

(2.) Differential Compaction within reef complex
(under overlying load, diff facies have diff. capacities for compaction.

(3) Fracturing - prob. largely caused by diff. compaction - tends to provide fair to good intercommunication thruout reservoir

(4) Effects of circulating waters

(5) Deformation from basement control.

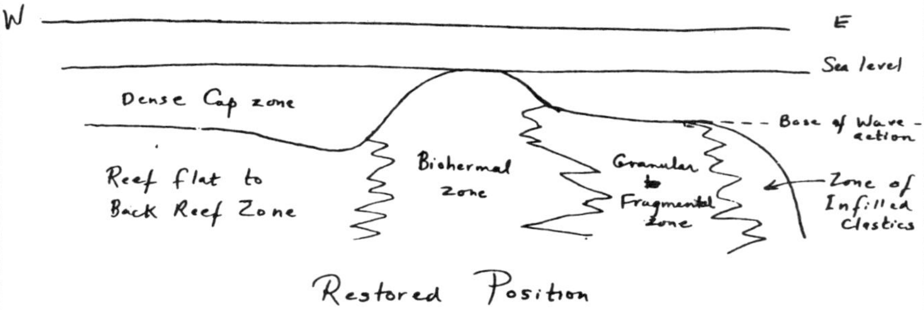
W
E
Sea level
Dense Cap zone
Base of Wave-action
Reef flat to Back Reef Zone
Biohermal zone
Granular to Fragmental Zone
Zone of Infilled Clastics
Restored Position

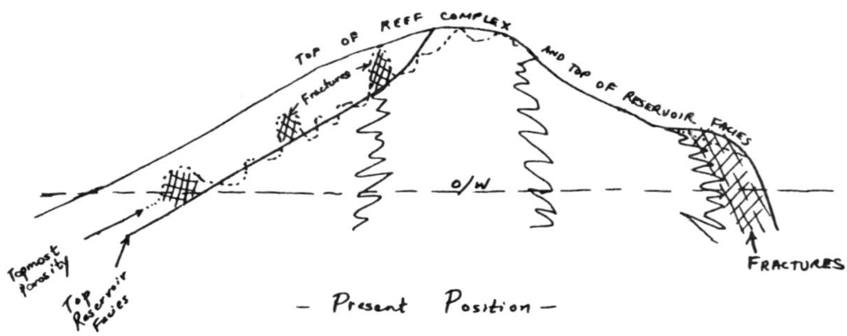
TOP OF REEF COMPLEX
AND TOP OF RESERVOIR FACIES
Fractures
O/W
Topmost porosity
Top Reservoir Facies
FRACTURES
– Present Position –

Appendix X

BAUGH'S RESERVOIR PARAMETERS

J. E. Ted Baugh, then with Sun Oil, published a paper in World Oil, Dec 1950, which preceded Haskett's. Here are some comments from his report:

Imperial-Simmons lease block is on one of the high spots of the reef.

Oil/water minus 1238 ft subsea

Extreme variation in "nature and degree" of porosity.

Gravity 35.5· API (34 to 36 API according to McCord)

Original reservoir pressure 1050 psi

Temperature: 94·F

Bubble Point: 485 psi

Gas/Oil Ratio: 180 (residual oil)

Gas/Oil Ratio: 161 (reservoir fluid basis)

Gas/oil Ratio: (McCourt) 155

Sulphur (crude) 0.5 to 1%

Sulphur (crude)(McCourt) 0.69%

Sulphur (gas) 3.5%

Shrinkage factor: 0.88

PI: 120 bbl/day per one-lb drop

PI: some as high as 400 bbl/day per one-lb drop in pressure

PI: average per net foot of pay 1.5

Flowing bottomhole pressure only a few pounds below static pressure

Acidizing doubles productivity index

Presence of at least a partial water drive

Pressure equalization is rapid and uniform.

World Oil, Dec 1950: p 260, James E. (Ted) Baugh . "Ted states that a considerable amount of data has been obtained ... The Devonian contains three main porous zones, the latter being the D-3 "recognized as being the local high, or a large coralline reef. He states that the oil-bearing section of the reef has a maximum thickness of about 200 ft. He also states that the average porosity is 8% (excessive) with maxes up to 15%, but it doesn't say how these values were determined.

Porosity occurs as large vugs and fine intergranular voids with "extremely good permeability". Redwater crude is fairly high in sulphur and chloride content, the former being 0.5%-1% while the H2S content of the gas is about 3.5%.

To indicate the high productivity of the wells - that even at high flow rates the flowing bottomhole presssure is only a few pounds per square inch below static pressure, approaching the precision of the pressure gauges in use.

Baugh estimates 490 barrels of oil-in-place per acre-foot.

Completion time is about 25 days (this was greatly reduced). Both open-hole (barefoot) and perforated liner completions are used

Oil Field Technical Society Papers: COGI, July 1950, Ball 3, No. 5. John R. McKay, then Division Petroleum Engineer with B-A, joined forces with Jack Henley, formerly Lane-Wells to discuss the neutron log. Its function is based on the concept that the neutron curve response is a function of formation fluid content. And the formation fluid content, in turn, is a function of porosity. The basic neutron curve evaluation for a cored well is carried out by plotting known core values against the neutron curve pen travel. And after going through various calculating hoops, they conclude that this method has " a very reasonable degree of accuracy". But it seems it seems to rely on tying in the log with the cores and their calulated porosities.

Looking back on the attempts of this curve to try to do some good, the authors conclude with seven points, the main one apparently being uniformity of logging speeds, same sensitivity, scale and bore hole size.

Repeat Paper printed in COGI by what appears to be a 1950 date

Appendix XI

FOURNIER'S RESERVOIR PARAMETERS

- 6 -

Redwater Oilfield

The following tabulation of pertinent data on the Redwater field was obtained from Imperial's Engineering Dept.

Productive Area	37,000 acres
Av. Depth to oil zone	3300 feet
Av. Effective Thickness of pay	94 feet
Max. " " " "	148 feet
Max. thickness pay	205 feet (Egremont No. 17)
Average porosity	6.2%
Average permeability	700-800 millidarcys
Water-oil contact	1236 feet subsea
Reservoir Temperature	94° F.
Original Reservoir pressure	1050 PSI
Saturation pressure	485 PSI
Gravity of Crude	34.8° API (sour)
Flash Original GOR	215 CF/B

Viscosity of oil at Saturation Conditions and 94°F - 2.5 centipoise

Shrinkage factor	.88
Original Recoverable oil	497,000,000 bbls.

The restored reef map of Redwater will prove of value to the geologists and petroleum engineers who are currently working on the zoning of the various coral and algal types across the reef from lagoon to outer edge.

F. L. Fournier

F. L. Fournier

Appendix XII

HASKETT PAPER, CIM, 1951

Reservoir Analysis of the Redwater Pool

*By I. HASKETT**

(*Transactions, Volume LIV,* 1951, *pp.* 140-150)

ABSTRACT

The Redwater pool, in north-central Alberta, was discovered in October, 1948, and at the end of 1950 was about 85 per cent developed. Production is from a Devonian coral limestone reef and is found at an average depth of 3,150 feet. The pool has an estimated productive area of 37,000 acres, and a maximum oil zone thickness of slightly over 200 feet. Wide variations exist in both type and values of porosity, with the average value determined for the reservoir being 6.16 per cent. The total stock tank oil in place is about 1.24 billion barrels. The crude has a saturation pressure of 485 psig, compared to the original value of 1,050 psig for the reservoir, and the principal recovery mechanism so far has been the natural water drive.

Cumulative production to January 1st, 1951, has been about 15.5 million barrels. These withdrawals resulted in a reservoir pressure decline of slightly over 100 psi. Predictions of future behaviour indicate that the water drive will not be of sufficient magnitude to maintain the reservoir above saturation pressure past 1953, and that, with natural depletion, the pool will be produced largely by a solution gas drive. Pressure maintenance by water injection has been considered and the possibility of raising the pressure to the original value of 1,050 psig is also discussed.

Introduction

WITH the discovery of the Leduc pool, early in 1947, the Devonian limestone became the major geological formation contributing to Western Canada's rapidly growing petroleum reserves. Redwater, discovered in October, 1948, is now the largest known Devonian pool in Canada, with a recoverable reserve estimated at approximately 500 million barrels.

Until recent years, the knowledge of the nature and characteristics of this type of limestone reservoir has been comparatively limited. For this reason and because of the size and importance of Redwater, the pool offers a challenge to petroleum technologists who are responsible for determining the development and producing practices which will yield the optimum ultimate recovery. Due to careful planning from the beginning, the amount of geological and reservoir data on Redwater is quite complete. The newest techniques of well logging, coring, and core analysis have been used and some improvements of standard techniques have been made during the development of Redwater itself. All operators in the pool have contributed to the accumulation of this information, most of which is released to all who have a direct interest in studying the Redwater reservoir. This paper reports on the results of one study of the pool dealing particularly with the nature of the recovery mechanism.

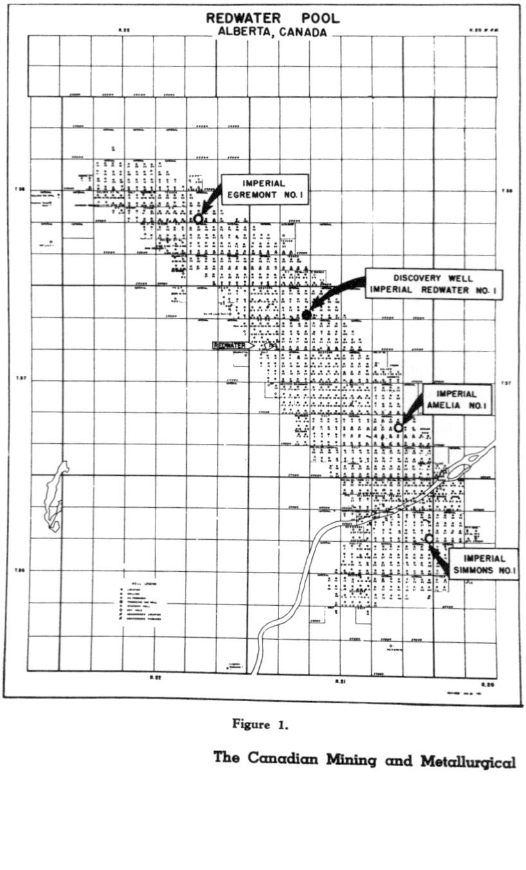

Figure 1.

* Imperial Oil Limited, Calgary, Alta.

History of Development

The Redwater field is in north-central Alberta, 35 miles northeast of the city of Edmonton. Drilling based on geophysical exploration led to the discovery of the pool by the Imperial Redwater No. 1 well, in Lsd. 1, Sec. 32, Twp. 57, Rge. 21, W4 (Figure 1). This well was drilled to a total depth of 3,264 feet, 151 feet below the top of porosity in the Leduc zone and 14 feet below the oil-water contact. Seven-inch casing was run to the top of the pay zone, and a five-inch liner cemented to bottom. The well was completed by gun perforating the interval 3,131 to 3,218 feet with 254 shots on October 12th, 1948. The initial yield, after a total acid treatment of 19,000 gallons, was 1,720 barrels per day of water-free 35° A.P.I. gravity oil, with a gas-oil ratio of approximately 190 cu. ft. per barrel.

The Imperial Egremont No. 1 well, in Lsd. 3-14-58-22, completed in December, 1948, and Imperial Amelia No. 1, in Lsd. 9-11-57-21, completed in January, 1949, determined the northwest-southeast strike of the field, and extended the productive area about 4½ miles in both directions from the discovery well. Also in January, production was obtained in Section 36, two miles to the west of Imperial Redwater No. 1. Development proceeded rapidly on a 40-acre spacing pattern and, by the end of 1949, 296 successful producing wells had been drilled. The Imperial Simmons No. 1 well, completed October 21st, 1949, extended the field approximately four miles farther to the southeast. In 1950, 457 producers were completed, making a total of 753 wells by the end of the year.

Except in the southwest and northwest corners, dry holes have fairly well defined the productive area, and it is expected that full development of approximately 900 wells will be reached in 1951. With the exception of a few edge wells, initial potentials have been well over 1,000 barrels per day after an acid treatment of usually from 1,000 to 5,000 gallons. Practically all wells have been open-hole completions.

Regional Geology

The production in the Redwater field is from the reefal Leduc (*D-3*) member of the Woodbend formation of Upper Devonian age. The reef is a coralline bioherm, about 900 feet thick, with an areal extent of approximately 200,000 acres. Situated on the eastern flank of the Alberta syncline, it is affected

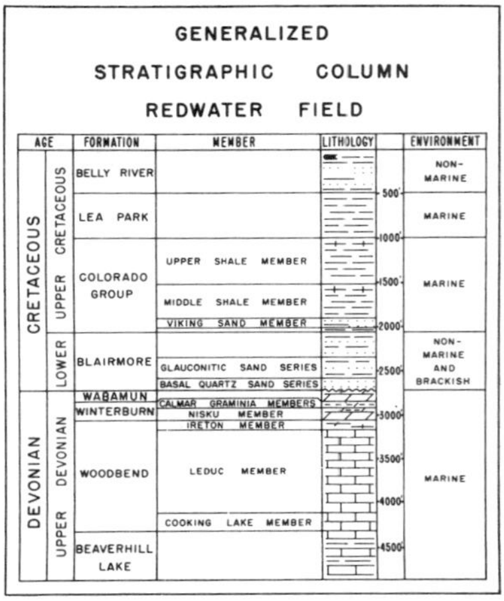

Figure 2.

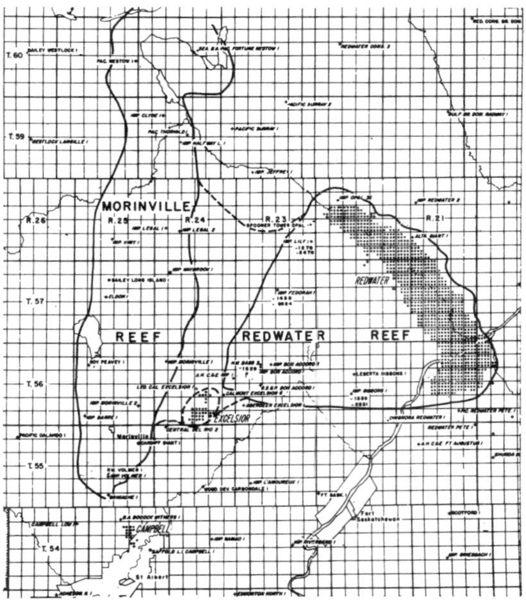

Figure 3.—Leduc (D-3) reefs, Redwater area.

by the regional dip to the southwest, and the productive zone occurs along the high northeastern rim.

Figure 2 is a generalized stratigraphic column for this area. The overlying Ireton member consists chiefly of green shale, and acts as the main cap-rock for the reef. The Cooking Lake member, also known as the fragmental limestone, is a blanket-type formation extending over a large part of Alberta, and has a thickness of from 200 to 250 feet. In areas of reef development, this zone has porosity and permeability characteristics similar to those of the Leduc, and becomes practically indistinguishable from that member.

The probable areal extent of the Redwater reef is shown in Figure 3. The south and northeast edges have been determined from seismic data, and additional control is provided by a number of dry holes. Along the northwest edge, however, geophysical indications are much less definite, and no wells have been drilled.

Several miles to the west is the Morinville reef, which is also built on the fragmental limestone to a thickness of about 900 feet. This reef is dolomitic, and is generally considered to be entirely separate from the Redwater reef. However, it is possible that a connection of either equal or reduced thickness is present in the intervening area. The Fedorah No. 1 well, in the Redwater reef, found about half limestone and half dolomite, making it appear that the differences in rock material in the two reefs might be the result of a gradual change from east to west. The Imperial Waybrook No. 1 well has been completed in the Morinville reef as a pressure observation well, and may eventually give some indication of the degree of continuity between the two reefs.

Reservoir Characteristics

A summary of basic data for the Redwater pool is given in Table I.

Production in the Redwater field is found at an average depth of 3,150 feet. The estimated productive area is about 37,000 acres, with the maximum thickness of the oil zone being slightly over 200 feet.

Over most of the pool area, a layer of non-porous, impermeable limestone is found at the top of the reef. This zone varies in thickness from 20 to 60 feet on the west side of the field, becomes generally thinner up-dip, and is negligible in some wells on the eastern flank. Since this material acts only as an additional cap-rock, all reservoir work has been based on the top of the porous section rather than on the top of the Leduc (*D-3*) formation. Figure 4 is a structural map of the reservoir, contoured on the tops of porosity as determined from cuttings, logs, and core analyses. A northeast-southwest cross-section of the field is shown with electric logs in Figure 5.

Estimates of the original water table in fifteen of the early wells drilled, made from electric logs, drill stem tests, and core data, varied from 1,234 to 1,241 feet sub-sea. The value of 1,236 feet sub-sea was accepted for use in all reservoir areas, volume, and behaviour calculations.

The reservoir rock is a light brown to cream-coloured limestone with only a small amount of dolomitization. In some parts of the field thin shale stringers are present, and dense limestone streaks having negligible porosity and permeability are scattered throughout the porous zone. Stylolites are common in both the dense and porous material. In the Simmons area, heavy secondary deposits of pyrobitumen are found. Most of the void space in the reservoir is made up of the vugs, which exist throughout the productive zone, and the porosity of the sand-like intergranular material, which is

TABLE I.— BASIC DATA FOR REDWATER POOL

Estimated productive area	37,000 acres
Water-oil contact	1,236 feet sub-sea
Average oil zone thickness	93 6 feet
Stock tank oil in place	1,242,700,000 barrels
Connate water (estimated)	15%
Shrinkage factor	0.883
Average porosity	6.16%
Original reservoir pressure	1,050 psig.
Saturation pressure	485 psig.
Reservoir temperature	94°F.
Solution G.O.R.— Flash liberation	215 cu. ft./S.T. barrel
Stock tank oil gravity	35° A.P.I.

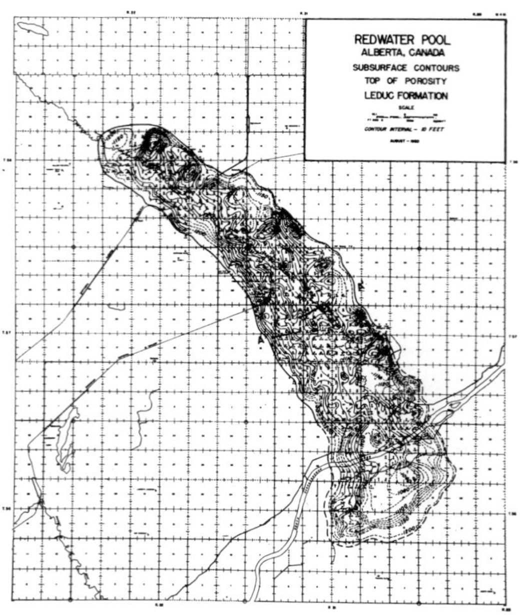

Figure 4.

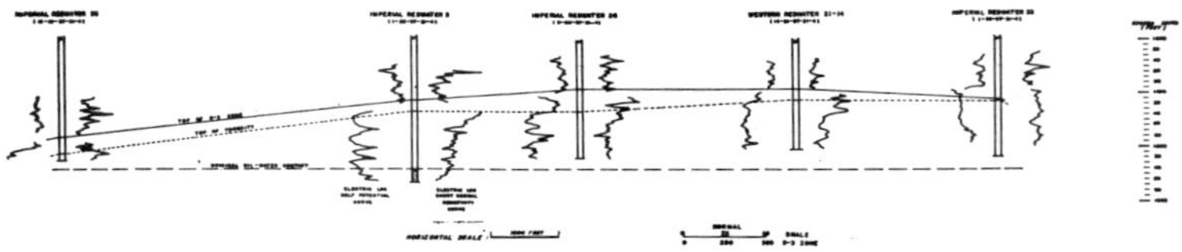

Figure 5.—N.E.-S.W. cross-section A-A[1], Redwater pool.

found chiefly along the edges of the reef. A smaller amount of porosity exists in the dense matrix and in fractures.

Diamond coring has been carried out on an extensive scale over the entire field. During the early stages of development, the minimum coring programme was a pattern of about four wells per section and, in some cases, all wells on a lease were cored. More recently, neutron-gamma ray logs have been substituted for cores on a large number of wells.

The heterogeneity of the reservoir rock has necessitated considerable deviation from typical core analysis procedure. All measurements are made on large size cores, as it was found that representative values were not obtained by the analysis of small plugs. At the time the samples to be tested are chosen, the footage for which each one is considered representative is also selected. In general, it has been found necessary to select and analyze an average of about one sample per foot of core to obtain good representation. In one case, the average porosity obtained by weighting the results of measurements on such samples was checked against the average porosity value determined by analysis of the complete core. Practically identical values were obtained, indicating that the porosities determined with this method of sample selection can be considered quite accurate.

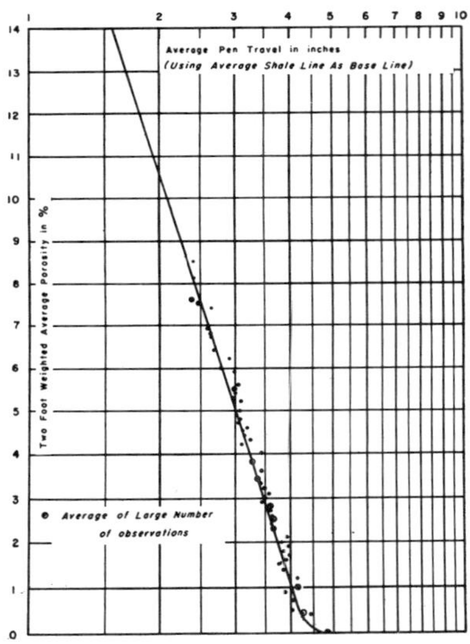

Standard deviation of major points = 0.43% porosity.
∴ Probable error in porosity = 0.25%.

Figure 6.—Correlation of neutron curve pen travel *vs.* porosity, Redwater pool.

In evaluating core analyses, only material having either a vertical or horizontal permeability greater than 1.0 md. has been considered effective. Porosity in less permeable sections is disregarded as probably containing a very high percentage of connate water, and as being essentially non-productive. The porosities determined for the effective sections are weighted over the to-

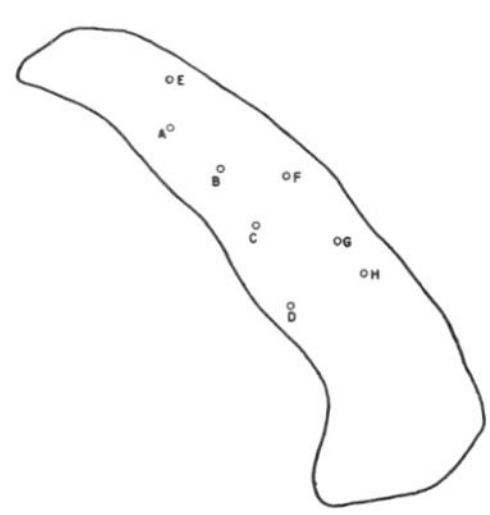

TABLE II.— TYPICAL CORE DATA, REDWATER POOL

WELL	A	B	C	D	E	F	G	H
Total oil zone thickness	89 ft.	126 ft.	55 ft.	59 ft.	91 ft.	130 ft.	97 ft.	98 ft.
Oil zone cored	71 ft.	112 ft.	37 ft.	41 ft.	84 ft.	106 ft.	92 ft.	98 ft.
Footage with permeability >1.0 md.	37 ft.	63 ft.	27 ft.	22 ft.	64 ft.	96 ft.	84 ft.	98 ft.
Average porosity of effective footage	7.1%	4.6%	7.9%	5.4%	15.4%	10.4%	10.6%	10.7%
Average porosity of oil zone	3.7%	2.6%	5.8%	2.9%	11.7%	9.4%	9.7%	10.7%

tal footage cored, and the value obtained is assumed to be the average porosity of the entire oil zone. Since trends of increasing or decreasing porosity with depth are apparent in some portions of the field, more accurate values are obtained when a high percentage of the section is cored.

In addition to the wide variations in cores found in individual wells, there is a definite trend across the field in the average oil zone porosity. On the west side, a large amount of interbedded dense material is present and the vuggy type of porosity, usually ranging from 4 to 10 per cent, is predominant in the effective portions. Toward the east, a much greater proportion of the zone is effective, and in some areas intergranular material, varying in porosity from 10 to over 20 per cent, predominates. The type of variations found, and the method of arriving at average porosity values, are shown for several wells on each side of the field in Table II.

Using approximately the same methods as those recently described by McKay and Henley (1), a correlation of porosity from core analysis and pen travel on neutron logs was made from ten wells distributed over the pool. The results are shown in Figure 6 and a detailed comparison of porosities obtained on the Imperial Egremont No. 32 well with the two methods is made in Figure 7. Average oil zone porosity values obtained from both core analysis and neutron logs usually check within a fraction of a per cent. Shale streaks may be detected on the gamma ray logs, and most of the other non-effective material may be eliminated by consideration of porosity values. In general, a porosity of from 0 to 2.5 per cent would be likely to have a permeability of less than 1.0 millidarcy.

(1) For references, see end of paper.

Figure 8 is an isoporosity map of the field. All available core analyses, neutron-gamma ray logs, and electric logs were used in constructing these contours. In general, the trends in porosity correlate very well with facies distribution maps which have been made, indicating further that the type of rock material present is the controlling factor.

By combining this map with the structural contours, an isopach map was obtained (Figure 9). The values shown represent the net fluid column thickness, as determined by multiplying the porosity at any point by the corresponding total thickness of the oil zone.

The effective permeability throughout most of the formation is quite high, probably due to the large amount of fracturing. Specific productivity indices of the wells range generally from 0.5 to 4.0 barrels/day/foot/psi. Except in the northwest portion, low pressure differentials also indicate good communication within the reservoir. The average effective horizontal permeability of the formation has been estimated as 500 millidarcies. While the effective vertical permeability is more difficult to determine, there is no indication that any dense limestone or shale streaks are continuous over large enough areas to be considered as major barriers to the vertical movement of fluids. Median ratios of horizontal to vertical permeability, determined for individual wells by core analysis, usually vary from 1.5 to 10.0. These values, however, probably do not give enough weight to vertical fracturing, since the core samples containing these fractures are often unsuitable for analysis

Reservoir Fluids

Bottom-hole samples from seven wells in the Redwater field have been obtained and analyzed. A summary of all available data is shown in Table III.

Reservoir Oil

The sample from Imperial Redwater No. 1 is considered to be generally representative of the reservoir fluid. The oil is undersaturated, having a bubble-point pressure of 485 psig at 94°F., compared to the original reservoir pressure of 1,050

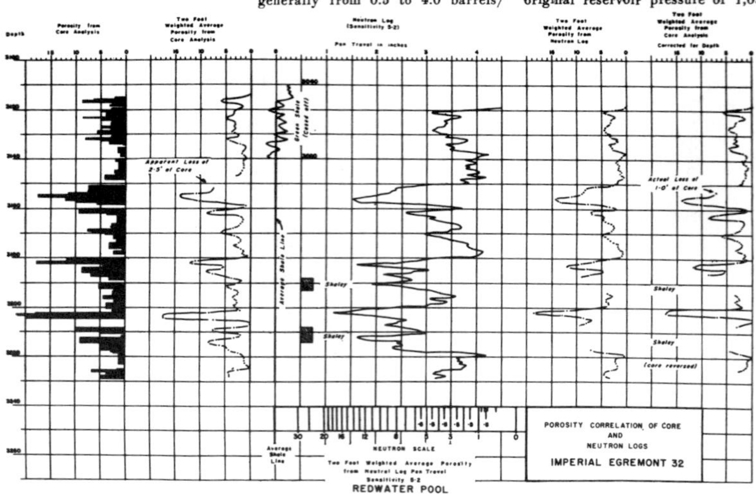

Figure 7.

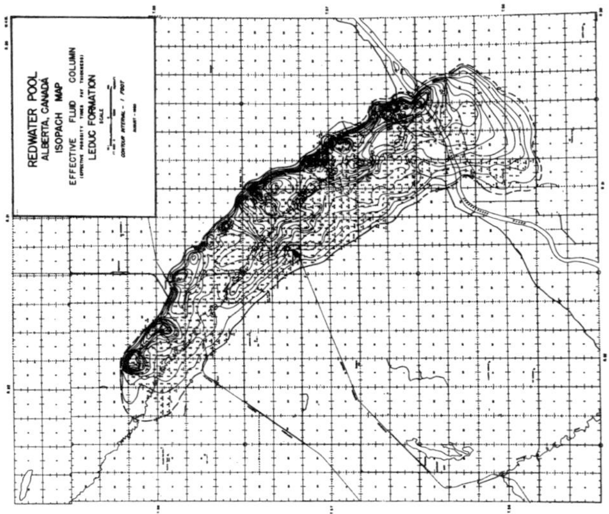

Figure 9.

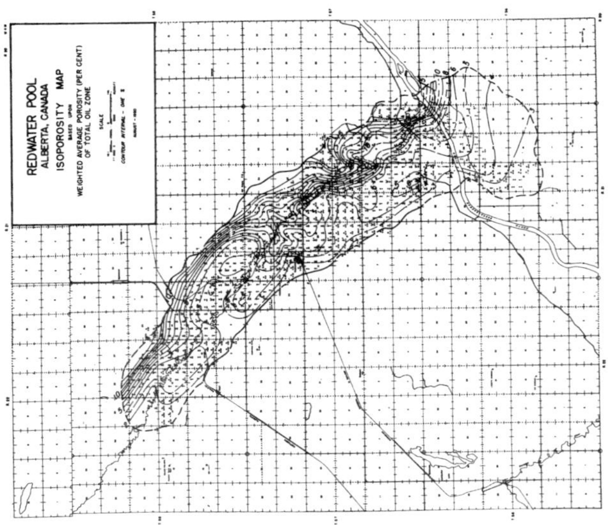

Figure 8.

Bulletin for April, 1951, Montreal

TABLE III. — SUBSURFACE SAMPLE AND RESERVOIR FLUID CHARACTERISTICS, REDWATER POOL

ORIGINAL RESERVOIR CONDITIONS
Pressure . . . . 1,050 psig.
Temperature . . . . 94°F.
Datum Depth . . . . 1,165 ft. sub-sea.

	IMPERIAL REDWATER No. 1	IMPERIAL AMELIA No. 1	IMPERIAL SIMMONS No. 1	ANGLO-HOME REDWATER No. 6	GULF SHKOLNY No. 16	TEXACO-McCOLL No. A-1	TEXACO-McCOLL No. B-1
SAMPLING CONDITIONS—							
Date	10-29-48	3-4-49	11-7-49	10-24-49			
Sampling depth–ft. sub-sea	1,128	1,138	1,135	1,035			
Sampling pressure–psig.	1,050	1,038	994	960			
Sample temperature—°F	94	94	94	93	96		
SAMPLE DATA—							
Saturation pressure—psig.	485	480	508	525	515	535	515
Compressibility — bbl/bbl/psi.	6.73×10^{-6}	7.94×10^{-6}	7.10×10^{-6}	7.7×10^{-6}	7.25×10^{-6}		
Specific volume at sat. pr. and temp.—cu. ft./lb.	0.02006	0.0200	0.0203	0.01984	0.02003		
Reservoir bbl./bbl. surface oil:							
Flash liberation	1.133	1.1468	1.1481	1.1307	1.1236		
Differential liberation	1.120	1.145	1.1175	1.1046	1.0924		
Solution gas—cu. ft./bbl. surface oil:							
Flash liberation	215.0	242.0	236.9	220.0	230.95	219	222
Differential liberation	180.0	211.0	186.1	181.0	173.73	178	187
Gravity of Stock Tank Oil:							
Flash liberation	34.8	33.8	33.6	34.2	33.7		
Viscosity of reservoir oil at 94°F. centipoises:							
At satur'n press.	2.68	2.50	2.52	2.46	2.53		
At sample press.	2.87	2.73	2.67	2.64	2.67		
SAMPLE COMPOSITION — Mol Per Cent							
Methane	10.54	9.06					
Ethane	6.20	6.37					
Propane	7.01	7.40					
Iso-butane	1.76	2.03					
N - butane	5.33	4.76					
Iso-pentane	2.80	2.33					
N - pentane	2.43	2.74					
Di-isopropyl plus		62.77					
Hexane	4.86						
Heptane plus	57.05						
Hydrogen sulphide	1.81	1.27					
Carbon dioxide	0.21	1.27					
Mol. weight—Residue	238	222					
Sp. Gr. —Residue	0.870	0.870					

psig at the datum plane of —1,165 feet. When flashed to atmospheric conditions, the fluid released 215.0 cu. ft. of gas per barrel of surface oil. The reservoir volume factor was 1.133 barrels per barrel of surface oil, and the liquid volume shrinkage was 11.7 per cent based on reservoir oil and 13.3 per cent based on surface oil. The gravity of the surface oil was 34.8° API.

By differential liberation. the fluid released 180 cu. ft. of gas per barrel of surface oil, and had a reservoir volume factor of 1.120 barrels per barrel of surface oil. The liquid volume shrinkage was 10.7 per cent based on reservoir oil, and 12.0 per cent based on surface oil. Figure 10*a* shows the gas solubility and shrinkage curves obtained by differential liberation.

The viscosity of the reservoir fluid, better represented by the Amelia No. 1 sample, varies from 2.73 centipoises at the original reservoir pressure to 2.50 centipoises at saturation pressure, and then rises rapidly as gas comes out of solution below this value (Figure 10*b*). The compressibility of the reservoir oil, determined by averaging data on all samples, is 7.34×10^{-6} bbl/bbl/psi.

Connate water

Up to the present time, no reliable average connate water value for the reservoir has been determined. A

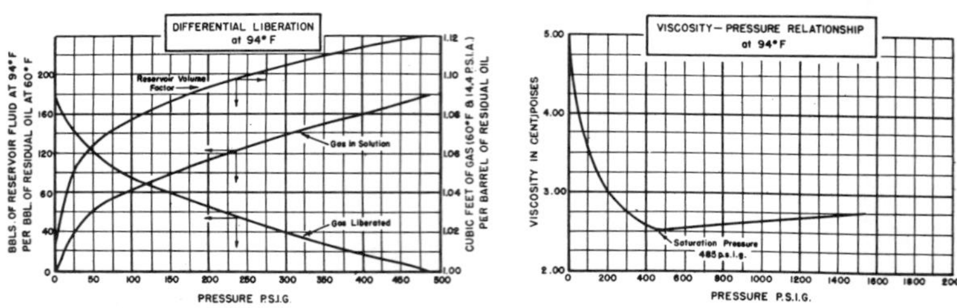

Figures 10*a* and 10*b*.—Fluid characteristics of reservoir oil, Redwater pool.

figure of 15 per cent has been used for oil in place calculations, this value being selected as typical of a vuggy limestone.

Results of distillation of cores taken with heavy oil, and laboratory mercury injection tests, indicate ranges of from 0 to 10 per cent connate water in large vugs, 20 to 30 per cent in intergranular material, and 50 to 100 per cent in the matrix porosity. Until a complete study is made to determine what proportion of the total void space in the reservoir is contributed by each type of porosity, there is no way to make an estimate of the weighted average value.

The relative amounts of void space in the vuggy material, which has low porosity but occupies the largest portion of the reservoir, and that in the smaller volume of high porosity intergranular material, will probably be the most important factors. Much of the matrix porosity is automatically eliminated due to low permeability, and that which is included because vugs or fractures provide permeability is only about 2 per cent porosity. Therefore, even with the very high connate water in this type of material, the effect it will have on the average for the reservoir is expected to be small.

Oil in Place

The total area included in the estimated productive outline of the field is about 37,000 acres. From the structural contour map, a total rock volume of 3,463,000 acre-feet was determined, and, from the isopach map, the total void space was found to be 213,390 acre-feet. The average oil zone thickness is, therefore, 93.6 feet and the average porosity is 6.16 per cent.

The original oil in place was calculated as follows:

Barrels per acre-foot....	7,758
Pore volume: Acre-feet	213,390
Barrels....	1,655,480,000
Hydrocarbon volume (connate water, 15%)	1,407,200,000
Stock tank oil in place (shrinkage, 0.883)	1,242,600,000

The pool has an average of 359 barrels of stock tank oil in place per acre-foot, and 33,600 barrels per acre.

A recovery efficiency of 40 per cent has been estimated. This may be somewhat conservative if the field is produced entirely by water, but very little is known about flushing efficiencies for this type of reservoir. Either laboratory flooding tests for the various types of porosity must be combined with a pore size distribution study, or the rise

Table IV.— Production and Pressure Data, Redwater Pool

Date	Total Wells Completed	Production Rate S.T. bbls/ day	Monthly Production S.T. Bbls	Cumulative Production S.T. Bbls	Reservoir Pressure End of Month Psig. at 1165 ft.
Original					1,050
1948—October	1	283	8,772	8,772	
November	1	371	11,137	19,909	
December	2	547	16,965	36,874	
1949—January	7	2,208	68,448	105,320	1,049
February	12	6,581	184,253	289,573	
March	19	9,702	300,770	590,343	1,043
April	41	11,451	343,540	933,883	
May	70	10,168	315,215	1,249,098	1,033
June	100	10,787	323,618	1,572,716	
July	126	12,797	396,714	1,969,430	1,024
August	158	17,740	549,947	2,519,377	
September	198	21,099	632,978	3,152,355	1,013
October	240	21,368	662,413	3,814,768	
November	274	21,053	631,590	4,446,358	1,006
December	296	12,388	384,007	4,830,366	
1950—January	313	14,483	448,978	5,279,344	1,007
February	334	19,564	547,783	5,827,127	1,005
March	366	28,132	872,088	6,699,215	
April	408	21,560	646,797	7,346,012	994
May	448	20,475	634,722	7,980,734	
June	521	26,145	784,362	8,765,095	981
July	573	28,383	879,863	9,644,958	
August	619	26,597	824,516	10,469,475	
September	654	36,593	1,097,773	11,567,248	
October	691	43,828	1,358,662	12,925,910	
November	725	50,573	1,517,207	14,443,114	951*
December	753	36,572	1,133,724	15,576,838	

*November 15th.

of the water table must be observed over a several year period before total recovery by water displacement can be accurately predicted. If the reservoir should fall below saturation pressure, a recovery of less than 40 per cent might be obtained with the resulting solution gas drive.

The total recoverable stock tank oil, using the above assumption, is 497,000,000 barrels, an average for the reservoir of 144 barrels per acre-foot. With the high effective permeability existing, it is probable that the water table will rise evenly throughout the field. As a consequence, drainage from high to low porosity areas will occur to the extent that all sections of the field will have an ultimate production of oil per acre-foot equal to the average value for the reservoir.

Pressure-Production History

By January 1st, 1951, a cumulative volume of 15,577,000 barrels of stock tank oil, or 1.25 per cent of the oil in place, had been produced. The most recently determined volumetric average pressure, for November 15th, 1950, was 951 psig, a decline of 99 pounds from the original value of 1050 psig. A summary of pressure-production statistics is given in Table IV and shown graphically in Figure 11. Although no accurate figures are available, the water production to date has comprised a negligible portion of the total withdrawals.

During the early life of the field, bottom-hole pressures were taken on an almost continuous basis, and no field-wide surveys were made in a short enough time period that the average reservoir pressure could be considered constant. This practice, and the large proportion of undeveloped acreage, made it impossible to construct accurate isobaric maps. To determine the average pressure decline of the reservoir, the field was divided into nine areas, and the volume of void space for each determined from the isopach map. All pressures measured were plotted by areas, and average pressure *versus* time curves were drawn. The early portion of the curves for areas in which there were no wells during the first year or year and a half were estimated from the pressure differentials apparently existing in various parts of the field, and from the first pressures measured when development was started. The average group pressures for various times were then taken from the curves, and were combined with the oil in place values to obtain the weighted volumetric average reservoir pressure.

With more complete development, it has been possible to construct isobaric maps, and the pressure distributions for July 1st, 1950 and November 15th, 1950, are shown in

Figure 12. These maps were made from field-wide pressure surveys taken over about one-month periods. All pressures were adjusted to the same date by using the average rate of decline for various portions of the field.

In general, good pressure communication has been evident throughout the reservoir. The first pressure measured for the Imperial Simmons No. 1 well, completed almost three miles from the nearest production in October, 1949, was 1,017 psig, a drop of 33 psi from the original value, and only 10 psi higher than pressures in the Amelia area at that time.

Evidence of poorer communication is shown by pressure measurements in the northwest portion of the field. There has been little tendency for these pressures to equalize either within this portion or with the remainder of the field, and the latest survey shows an increase in the pressure differentials. This behaviour may be due to the incomplete development in this area and to the possibility that a smaller proportion of the oil in place is being withdrawn, with the lesser permeability preventing drainage to lower pressure areas.

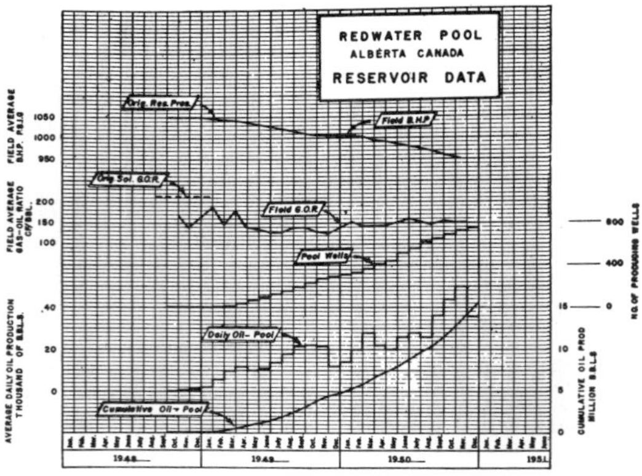

Figure 11.

Recovery Mechanism

A natural water drive has been the predominant recovery mechanism for the field to date, with about 94 per cent of the cumulative withdrawals having been replaced by the water influx. The average production per pound drop in pressure to the end of 1950 has been about 138,000 stock tank barrels, or 156,000 barrels of reservoir volume. The expansibility of the total oil volume, determined from the compressibility of the reservoir fluid, is only about 10,000 barrels per pound drop in pressure.

The principal source of the water drive is the expansion of water in the Redwater reef and the fragmental limestone immediately under the reef, these two zones having an average total thickness of about 1,100 feet. Some additional energy is possibly supplied by expansion of the formation rock, and compaction of the overlying shale.

The fragmental limestone has been found to be porous and permeable in about 50 per cent of the wells drilled outside of areas where *D*-3 reefs are present. Therefore, limited pressure communication probably exists throughout this zone, and later in the life of the pool this formation would be expected to contribute to the water

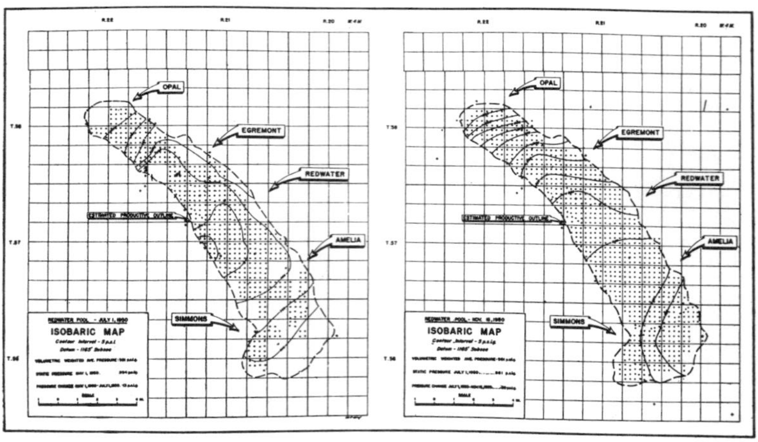

Figure 12.

drive at Redwater. The Morinville reef might also have some effect due to the probable connection through the fragmental limestone, even though no reefal connection exists. However, the large distance of this reef from the productive zone would delay and minimize any water drive from this source. Pressures measured in the Waybrook No. 1 well should aid in future determinations of the effect of this reef on Redwater.

Several aspects of the water drive at Redwater are unusual. The area of the pool is almost one-fifth of the total aquifer area, if only the main Redwater reef is considered, but the average oil zone thickness is less than one-tenth the total formation thickness. Therefore, not only is the pool completely underlain by water, but about 20 per cent of the total volume of water available for expansion lies directly beneath the oil zone. In addition, the shape of the reef presents a unique problem in the analysis of water flow from aquifer to pool. While it may be considered as a radial flow problem, by comparing the reef to a section of a cylinder, it is one of diverging flow, with the productive zone being along the outer circumference, and the volume of the aquifer decreasing with distance from the pool. In most water drive fields, the flow is convergent.

Although an accurate prediction of the behaviour of any water drive pool is a difficult mathematical problem involving the unsteady state equations of fluid flow, it is possible from simplified calculations to obtain some indication of the magnitude of the water drive to be expected.

Assuming the same porosity as exists in the field, the total volume of water in the Redwater reef is approximately 100 billion barrels. With an estimated compressibility of brine of 3.0×10^{-6} bbl/bbl/psi, the total expansibility of this volume is about 300,000 barrels per pound drop in pressure. Therefore, if the entire reef dropped uniformly in pressure, the total expansion from the original pressure of 1,050 psig to the saturation pressure of the oil, 485 psig, would be only about 170 million barrels, or enough to produce less than one-third of the recoverable oil in the field. In most cases, however, it has been found that an apparent compressibility considerably larger than the value for water must be used in order to explain the behaviour of an aquifer, probably due to other factors such as rock expansion and shale compaction. Therefore, if the pool were produced at low enough rates to allow time for almost complete pressure equalization throughout the Redwater reef, and to obtain a maximum effect from the fragmental limestone and the Morinville reef, it might be possible to produce the major portion of the reserves with the natural water drive.

At any reasonable and economic rates of production, however, a pressure sink will occur in the productive area, and a large pressure differential would be expected to develop across the reef. The resulting water influx is then only a portion of that which would occur if the total amount of energy available from expansion in the aquifer were utilized.

The past pressure-production history of Redwater may be used to obtain a better general indication of future behaviour. With production rates which have been very low for the size of the reserve, the water drive has been sufficient to produce an average of 138,000 barrels per psi drop with a maximum of 164,000 bbls/psi being reached in the period between the last two pressure surveys. Assuming the latter figure for the future, the reservoir would drop to saturation pressure after a cumulative production of about 93 million stock tank barrels of oil, or 7.5 per cent of the oil in place. This may be considered a minimum value, since, with a further decline in reservoir pressure, greater differentials would develop across the reef and an increase in the rate of water influx would be anticipated.

From the above calculations, it may be concluded that the natural water drive will not be of a sufficient magnitude to maintain the reservoir above saturation pressure, and that, unless a programme of artificial pressure maintenance is initiated, the pool will be produced principally by a solution gas drive. Increasing gas-oil ratios, high drawdown pressures, and excessive water coning would then characterize individual well production, and the overall recovery efficiency would be further reduced by shrinkage and by a large increase in the viscosity of the reservoir fluid. Depletion in this manner would, therefore, be expected to result in a much lower ultimate recovery.

Future Reservoir Behaviour

Production rates which have been assumed for the future may be summarized as follows:

	1951	1952-1954
Av. prod'n rate, S.T. bbls/day..	65,800	81,300
Total Yearly Prod'n, S.T. bbls..........	24,000,000	29,700,000
Max. well prod'n rate S.T. bbl/day..........	93	111
Field deplet'n rate, %/year..	4.8	6.0

These rates are much higher than any at which the pool has yet been produced. However, they do not exceed a depletion rate which is considered efficient for a pool of this type, if the reservoir is maintained above the saturation pressure by either a natural water drive or artificial pressure maintenance. Nor are individual well rates high enough to cause damage due to high pressure draw-downs and excess water production due to coning.

In order to predict accurately the effect of these withdrawal rates on the average reservoir pressure, a study was made with the Carter oil pool analyzer (2). This instrument, developed especially for analysis of water drive reservoirs, makes use of an analogy between electrical and fluid flow equations. Electrical resistance and capacitance units are used to represent the flow resistance and expansibility of various portions of the formation, and any aquifer configuration may be considered by a proper choice of values. Voltage and currents throughout the circuit are proportional to pressure and rates of flow in the fluid system. Trial and error adjustments are made until the voltage of the unit representing the pool follows the pressure history of the pool when currents proportional to past production rates are withdrawn. Predictions are then made by recording the voltage decline when assumed future production rates are considered.

Natural Depletion

A prediction of the natural pressure decline for the Redwater pool is shown in Figure 13, together with the monthly reservoir withdrawal rates which have been assumed. With only the natural water drive as the recovery mechanism, the pressure at the datum plane would fall to 530 psig by the end of 1953. At this time, the fluid at the top of the producing zone, some 125 feet higher, would be at the saturation pressure, and any further pressure decline would result in the release of gas from solution. Production rates would then have to be decreased in order to utilize to the greatest advantage the partial nat-

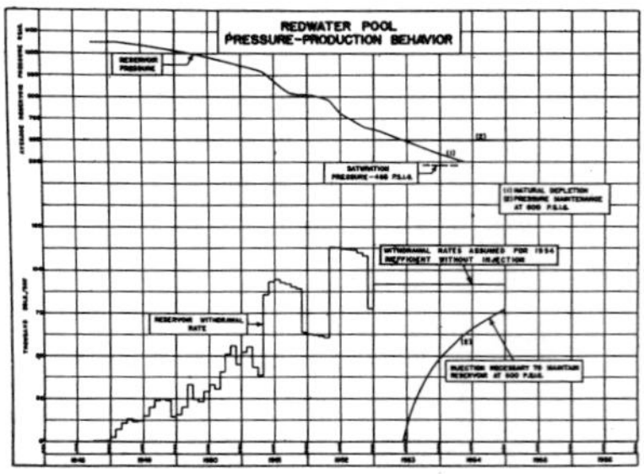

Figure 13.

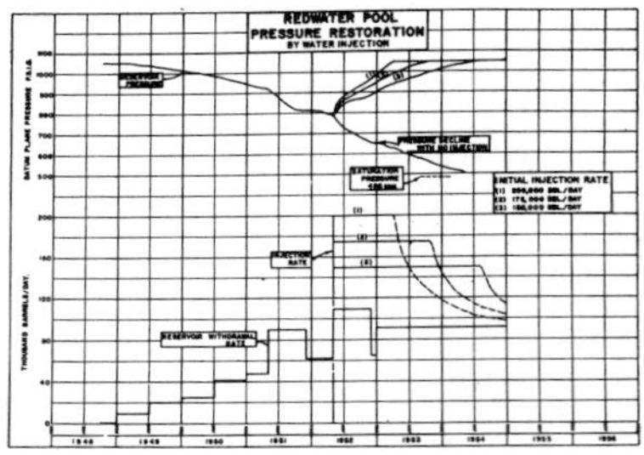

Figure 14.

ural water drive, and to obtain as large an ultimate recovery as possible with the solution gas drive which would result.

Pressure Maintenance

A greater ultimate recovery will be obtained, and the higher rates of production can be continued, if an artificial pressure maintenance system is installed to prevent the reservoir from dropping below the saturation pressure. The conditions existing in the aquifer are ideal for the injection of water, either directly beneath the field, considerably below the water-oil contact, or into the aquifer along the west edge of the pool. The North Saskatchewan river, which flows through the field, provides a convenient source of fresh water, and it is not anticipated that any serious treating problems would be encountered in the injection of this water.

The rates of injection necessary to maintain the reservoir pressure at 600 psig up to the end of 1954 are also shown in Figure 13. After this time, the proportion of the withdrawals which would have to be replaced with injection would continue to increase, due to a diminishing rate of the natural water influx. Eventually, practically complete pressure equalization would occur throughout the aquifer, and all net withdrawals from the reservoir would have to be replaced by injection.

Pressure Restoration

Consideration has been given to the possibility of injecting enough water into the formation to raise the reservoir pressure to the original value of 1,050 psig. This procedure would not be expected to result in any greater ultimate recovery than would be obtained with pressure maintenance at 600 psig. However, practically all wells in the field except those making water could be produced by natural flow, and considerable savings in operating expense would result. The increase in the volume of water which would enter the reservoir, compared to pressure maintenance at 600 psig, would only be the amount that the undersaturated oil in the reservoir would be compressed. It has been calculated that the additional rise in the water table would be only about one foot, causing no significant reduction in the producing life of individual wells.

Predictions of the pressure rise for various rates of injection, assumed to start in May, 1952, are shown in Figure 14. A constant rate of injection of 150,000 barrels per day would raise the pressure from 790 psig to 1,050 psig in 27 months, 175,000 bbl/day in 18 months, and 200,000 bbl/day in 11 months. During the period of rising pressure, and for some time after the original value was reached, the natural water drive situation would be reversed, with the pressure near the field being higher than the remainder of the reef. Injection rates could be gradually diminished after the desired pressure was attained, but would have to be somewhat larger than the net withdrawal rates until the entire reef were again equalized at the original pressure. This should occur when the cumulative volumes of injection and withdrawal become equal. Then, as in the case of pressure maintenance at the saturation pressure, injection would only have to be sufficient to replace withdrawals.

Acknowledgements

The writer wishes to express appreciation to the Management of the Western Producing Division of Imperial Oil, Limited, for permission to publish this paper, and to Mr. Robert Pot and other members of the petroleum engineering and geological staffs who have materially assisted in its preparation.

References

(1) McKay. John R., and Henley, Jack. *A Practical Evaluation of the Neutron Log for Canadian Oil Fields;* Can. Oil and Gas Industries, Vol. 3, No. 5, July, 1950, pp. 50-53.

(2) **Bruce, W. A.,** ***An Electrical Device for Analyzing Oil-Reservoir Behavior;*** A.I.M.E., Trans., Vol. **151, 1943. pp. 112-124.**

Appendix XIII

AUTHOR'S DISSERTATION ON THE GREEN HORNET REPORT

The Redwater reef is saucer-shaped and, when it was tilted to the northeast, the oil migrated up-dip. It is complex because the types of rock are controlled by the different environments or zones present during carbonate build-up. Five zones are identified.

1. Biohermal - major organic growth facing onto the sea (heterogeneous);
2. Granular fragmental zone - formed by fragments broken from the reef body;
3. Infilled clastic zone - finer granular fragmental lime plus precipitated lime resulting in destruction of porosity;
4. Reef-flat to back-reef zone (west side of the productive zone) - the site of physico-chemical deposition of fine lime. The rock tends to be bedded, or biostromal;
5. Dense cap zone - created after organic reef building ceased (virtually non-porous).

The next part of the report discusses how the top of the D-3 was determined using E-logs and radioactive logs on approximately ten wells. Gunnar Haugrud recalls that his fellow workers would argue over six inches as to the top of porosity - "no consistent tool everywhere available." The report discusses limitations in the choice of picking tops, which is beyond the purview of this book.

Core analysis was given the most weight when the well was cored, but this did not always work where there was poor and/or streaky porosity.

The next item was water-oil contact. The original water table was taken at

-1237 feet (subsea). Calculations - assuming the average field porosity was 6.53 per cent, connate water of 15 per cent, and cumulative production of 86.4 million stock tank barrels - would show a rise of between 10 and 12 feet. "It has been assumed that the water table has risen evenly over the entire field ... There are indications this might be the case but insufficient data are available to accurately determine any variations."

The next division contemplates porosity and how it was estimated. Again, this detailed exercise is beyond the limits of the book, but it does indicate the exhaustive studies that were undertaken to arrive at a realistic set of values.

The report cautions: "Certain inconsistencies in the data have been averaged out, and in the few isolated cases where cored wells have had porosities considerably higher or lower than the average for the zone, these inconsistencies have been disregarded. Futhermore, values assigned to broad bands of porosity do not necessarily apply to individual locations. Variations as great as 2%-3% are known to exist for four or five wells."

Despite these heroic efforts on the part of the Working Committee, one can understand why the Operators' Committee would have misgivings about an equitable unitization formula. Is this why they fell back on production? And yet there was, "...Insufficient data to determine why certain wells were underproduced. And while most of the wells which are making water are included as being under-produced, some wells are producing relatively large amounts of water, but are still making their allowables. This raises yet another concern about using production as the benchmark."

The textual part of the report is then followed by a number of plats and graphs which confirm the impossibility of reaching an equitable unitization formula.

Sample Pages from

THE GREEN HORNET REPORT

SECTION 3

CORE ANALYSIS SUMMARY

Location	Well	Date of Anal.	Laboratory	Samples Anal. Per Foot of Core	Total Pay Thick-ness Feet	Footage Not Cored Top Feet	Footage Not Cored Bottom Feet	Core Re-covery Percent	Amount of Rec. Core Tested Percent	Porosity of Eff. Footage $K > 0.1$ md Percent	Effective Footage $K > 0.1$ md Percent	Wtd. Avg. Porosity $K > 0.1$ md Percent	Calc. or Adj. Values Key Wells Percent
T.58-R22 Sec. 16	McMahon #16-16	6-50	Core L.	1.00	90	4	40		100	3.77	98	3.69	xxx
	Texaco # D-1	5-50	"	1.00	58		20	99.5	100	3.06	89.7	2.74	2.74
	" # D-3	6-50	"	1.03	61	6	23	96.5	100	7.43	82.5	6.13	xxx
" 15	" # C-1	11-49	Chem. & Geol.	2.95	129	1	33	98.2	98	5.87	53.1	3.12	(4-5)
	" # C-2	12-49	"	1.00	72		26	39.2	100	8.49	100	8.49	xxx
	" # C-3	4-50	"	1.00	81		30	74.5	100	4.50	78.9	3.55	xxx
	" # C-4	4-50	"	1.00	99		24	97.3	100	6.37	67.1	4.27	4.27
	" # C-5	5-50	"	1.00	92		23	95.8	100	8.66	83.8	7.26	(7-9)
	" # C-7	5-50	"	1.00	76		22	94.5	100	5.43	81.4	4.42	4.42
	" # C-8	5-50	"	1.00	72		24	92.8	100	5.76	77.5	4.46	4.46
	B. A. #3-15	3-50	Core L.	.74	59	1	29	91	100	3.84	92	3.55	(2-4)
	" #7-15	3-50	"	.85	83		35	89	100	3.38	100	3.38	(4-5)
" 14	Imp.Egremont #1	1-49	Madison	.84	83	13		100	85	5.69	100	5.69	5.69
	" " #60		Core L.	.99	86	1		96	100	13.64	93	12.67	12.67
" 12	" " #71		"	.96	153	5		100	100	4.27	57	2.42	2.42
" 11	" " #14	4-50	Chem.&Geol.	1.20	87		19	100	100	6.65	62	4.11	4.11

Well		No.	Location	K.B. Elev. Feet	Top of D-3 Feet	D-3 Sub-sea Depth Feet	Top of Porosity Feet	Porosity Sub-sea Depth Feet	Original Pay Thickness Feet	Adjusted Pay Thickness Jan. 1/54 50% Rec. Feet	Adjusted Pay Thickness Jan. 1/54 60% Rec. Feet	Porosity Band PerCent	Cumulative Prod. Dec.31/52 Bbls	Prod. Rate Dec. /52 Oil B/Mo.	Prod. Rate Dec. /52 Water B/Mo.	Prod. Rate Dec. /52 Water %	Remarks
Imperial	Amelia	#7	5-14-57-21	1986	3080	1094	3082	1096	141	129	131	10	89,279	2446	8		
"	"	#8	16-10-57-21	1987	3078	1091	3090	1103	134	122	124	5	79,609	2212	4		
"	"	#9	4-14-57-21	1992	3090	1098	3093	1101	136	124	126	10	104,773	2291	10		
"	"	#10	12-10-57-21	2026	3124	1098	3154	1128	109	97	99	5	86,939	2251	2		
"	"	#11	5-10-57-21	2058	3172	1114	3194	1136	101	89	91	5	89,378	2353	6		
"	"	#12	4-10-57-21	2052	3164	1112	3183	1131	106	94	96	5	85,862	2361	6		
"	"	#13	3-10-57-21	2050	3148	1098	3165	1115	122	110	112	5	83,309	1935	4		
"	"	#14	2-10-57-21	2012	3102	1090	3116	1104	133	121	123	5	87,793	2316	6		
"	"	#15	16-4- 57-21	2054	3180	1126	3205	1151	86	74	76	5	89,472	2428	6		
"	"	#16	9-4-57-21	2062	3183	1121	3209	1147	90	78	80	5	81,075	2296	2		
"	"	#17	3-14-57-21	2002	3125*	1123	3128	1126	111	99	101	10	93,398	2285	7		
"	"	#18	6-14-57-21	2062	3140	1078	3142	1080	157	145	147	10	80,350	2172	7		
"	"	#19	11-14-57-21	2070	3177	1107	3178	1108	129	117	119	10	68,449	2180	7		
"	"	#20	14-14-57-21	2051	3197	1146	3199	1148	89	77	79	10	75,545	2255	581	20.5	(LW)
"	"	#21	6-10-57-21	2019	3115	1096	3129	1110	127	115	117	5	71,284	2297	6		
"	"	#22	11-10-57-21	2018	3119	1101	3153	1135	102	90	92	5	73,439	2219	4		(LW)
"	"	#23	5- 3-57-21	2049	3178	1129	3203	1154	83	71	73	5	80,765	2286	5		
"	"	#24	6-3-57-21	2043	3158	1115	3177	1134	103	91	93	5	78,502	2349	2		
"	"	#25	7-3-57-21	2036	3130	1094	3152	1116	121	109	111	5	84,100	2260	5		
"	"	#26	8-3-57-21	2039	3135	1096	3161	1122	115	103	105	5	77,074	2291	2		

SUMMARY 1

In December, 1952, a committee was organized to study certain reservoir characteristics of the Redwater Field. This report includes the basic reservoir data that might be considered in a participation formula if the Operators decide that the field should be unitized in order to initiate a pressure maintenance program. Using all geological and well information available, the top of porosity for each well, the original water-oil contact, the rise of water for two possible recovery factors and the porosity variations in the field were determined. For the calculation of oil in place and reserves, the connate water and recovery factors used were assumed overall averages for the field. Inasmuch as there could be considerable variation in these factors for the different types of porosity present, oil in place and recoverable reserves were not computed for individual wells. A summary of the basic reservoir data is shown below:

Number of Wells	925
Average LSD Area	40.67 acres
Original Water-Oil Contact (Subsea)	1237 feet
Average Pay Thickness (Original)	101 feet
Average Field Porosity	6.53 %
Estimated Average Connate Water	15 %
Shrinkage Factor	.883
Total Rock Volume	3,791,095 acre ft.
Original Stock Tank Oil in Place	1,441,475,000 bbls
Estimated Cumulative Production to Jan. 1/54	86,454,000 bbls
Recoverable Stock Tank Oil - 50% Recovery Efficiency	720,738,000 bbls
Recoverable Stock Tank Oil - 60% Recovery Efficiency	864,885,000 bbls
Oil-Water Contact Jan. 1/54 - 50% Recovery Efficiency (Subsea)	1,225 feet
Oil-Water Contact Jan. 1/54 - 60% Recovery Efficiency (Subsea)	1,227 feet

Original pay thickness and porosity maps as well as a tabulation of individual well data are included in the Appendix.

Appendix XIV

GAS PLANT START-UP PERSONNEL 1956

Altman, Hans	Steam Eng.
Ashton, Mr.	Operator
Bottolfs, Alf	Operator
Bryant, Glen	Gas Gathering
Cooper, Herb	Assistant Foreman
Erb, Dick	Steam Eng.
Gillespie, Ray	Trainee
Hanington, Mark	Steam Engineer
Hildabrandt, Oscar	Chief Steam Eng.
Martin, Armand	Gas Gathering
Maydanski, Marash	Operator
Melnyk, Walter	Operator
Ory, John	Operator
Petruk, Mike	Gas Gathering
Snider, Ed	Mechanic
Thomas, Sy	Foreman
Trelford, Eldon	Operator
Wintoniak, John	Loading Rack
Zolkewski, Phil	Operator

Appendix XV

IMPERIAL OIL REDWATER WRITE-UP

Foreword

Here is an Imperial Oil write-up circa 1970 which capsulates the chain of events from discovery to 1970. It serves as an informative window on those days.

- 2 -

PART A

POOL HISTORY

The Redwater D-3 pool was discovered in October, 1948. By early 1952, 925 wells had been drilled on 40 acre spacing. The productive section of the Devonian reef is found at an average depth of 3100 feet and had an original average pay thickness of 100 feet. The pool has excellent reservoir characteristics and has a strong water drive. The crude is 35° API gravity and contains 0.7 percent sulphur. The original oil in place is 1.3 billion barrels with recognized recoverable reserves of 780 million barrels. Cumulative oil production to the end of 1970 was 397 MMB and cumulative water production was 207 MMB.

Figure No. 1 is a graph of the pool production history showing the annual daily average oil and water production. The pool daily rates rapidly built up to a peak of 79 MBOPD in 1955 then declined to a low of 35 MBOPD in 1960 and has increased to 58 MBOPD in 1970. Water production in the field reached a peak rate of 56 MBWPD in 1964 and declined to the present rate of 30 MBWPD. The decline in water production was due to the shutting in of marginal wells with the change in production regulations.

All wells in the field are pumped. Conventional pumping unit rates vary from 100-1100 BFPD and electrical submersible pumping rates change from 2000-3500 BFPD.

The original field facilities consisted of conventional batteries where the produced fluids were collected, separated and measured. The oil was delivered to the Interprovincial Pipeline Terminal through two oil gathering systems in the field. The Imperial Pipeline Company System gathered about 80 percent of the oil and the British American Alberta Pipeline System about percent.

In 1950 the Redwater Salt Water Disposal Company Limited was formed to collect and dispose of produced water. The system was designed to operate

- 3 -

on gravity. Facilities were installed and seven wells capable of injecting 100 MBWPD were completed. The Disposal Company is owned by the field operators and is operated by Rice Engineering and Operating Ltd.

In 1955 Imperial installed the Oval Gas Gathering System and a 10 MMcf/D .s plant known as The Redwater Gas Conservation Plant. Gas was collected from the field batteries and processed in the gas plant. The residue gas was sold to Midwestern Industrial Gas Ltd. for delivery to Fort Saskatchewan. The propane and butane products are marketed through Edmonton and the pentanes plus are blended and sold with the crude oil.

In 1965 the Oil and Gas Conservation Board of Alberta changed the production regulations by introducing the Production Spacing Unit and Block concepts. This had a dramatic effect on production operations as it allowed the transfer of well allowables under certain conditions. Imperial benefited more than other companies as its acreage is contiguous, which permitted extensive grouping of producing wells and field facilities. The number of Imperial operated wells was reduced from about 500 to 119, and the number of batteries reduced from 32 to 18. Figure No. 2 is the present configuration of PSU's and Blocks in the Redwater Pool.

In 1968 Imperial further optimized the field operation by converting the oil and gas gathering systems to a two phase gathering system and installing a fieldgate battery to replace the conventional batteries. Field satellites were installed for well testing purposes.

Imperial operates these facilities and charges other operators a fee for collecting, treating and delivering crude to the Interprovincial Pipeline rminal at Redwater. A fee of four cents per barrel is charged for oil collected in the two phase system. The capacity of the two phase system is 80 MBFPD and the fieldgate battery 75 MBFPD. A dead oil gathering system collects dead oil from a few conventional batteries on the pool fringes and delivers the

- 4 -

crude directly to the terminal for a fee of five cents per barrel. The capacity of this system is 8-10 MBOPD. Imperial owns 81 percent and Gulf Oil 19 percent of the two phase system and fieldgate battery.

In August, 1971 the gas plant was expanded to 22 MMcf/d. Based on a fi average GOR of 187 cubic feet per barrel these facilities are equivalent to 120 MBOPD of oil production. In addition to the gas plant, two small field compressor booster stations collect gas from dead oil batteries and compress it for delivery to the main plant. Imperial purchases the solution gas for 1 cent per mcf and owns 100 percent of the gas processing and compression facilities.

Imperial owns and maintains three injection systems; (1) Imperial Redwater 39 (1-29-57-21), which is used for injecting plant process water into the reef aquifer, (2) Imperial Redwater 65 (7-28-57-21) which is maintained as an emergency condensate injection well and is completed in the D-3 oil zone, and 3) Imperial Opal 45 and 46 (2 and 3-29-58-22) which are completed in the D-1 reservoir, and used for storage of raw D-3 solution gas.

The pool productivity is now about 80 MBOPD. The average production rate in 1970 was 58 MBOPD and the peak day rate during the year was 84 MBOPD. The monthly average peak rate over the past 10 years has been 30 percent over the average annual rate. Similar fluctuations can be expected for the next few years.

Appendix XVI

DEED OF SURRENDER

(The First Canadian Crown Corporation)

Comments by Author

It was twelve years after Oliver Cromwell's death, and only ten years after Charles II ascended the throne, marking the restoration of the monarchy. On May 2, 1670, the monarch granted to the Hudson's Bay Company millions of acres best described as "those lands draining into Hudson Bay". It included what is now northern Ontario, northern Quebec, all of Manitoba, most of Saskatchewan, southern Alberta, and a portion of the present Northwest Territories. It was named Rupert's Land in honour of the king's cousin, who was appointed the company's first governor.

The governors knew what they were doing because 100 years later the rival North West Company was formed in 1779 by a group of Scots including James McGill, the Frobisher Brothers, and the Ellices with Peter Pond as their agent in the Athabaska country. Nevertheless, the indefeasible title which the Hudson Bay Company was granted assured it of its position across those vast areas where furs and liquor traded.

By the 1850s, Rupert's Land was being looked at as an annexation to what would be Canada. Indeed, provision was made in the British North America Act for this, as well as taking over the North-Western Territory.

The enabling legislation was called the Rupert's Land Act 1868. This required the "approval" of the new Government of Canada as represented by Sir George Etienne Cartier, a knighted member of the Canadian government. This arrangement was finally agreed upon in 1869 and included the transfer of the so-called North-Western Territory, along with the acquisition of the Hudson Bay Company's land as one package.

In the original letters patent granted by Charles II, the draftsmen of the agreement very cleverly referred to the outer limits of the area as being Hudson's Straits instead of Hudson Bay (present spelling) so nothing would be left out. Thus it was that the northernmost part of Quebec was originally Hudson's Bay land.

To make doubly sure that everything was included, the agreement referred to "all the land and territories upon the countries, coasts, and confines of the seas, bays, lakes, rivers, creeks, and sounds to be referred to as reckoned and reputed as one of His Majesty's plantations or colonies in America called Rupert's Land."

As with any deal, money is the first consideration. This was looked after in the negotiations between Canada and Her Majesty's[1] Government. The Imperial Government (not the oil company) made a promise: "Guarantee be given to a loan of £300,000 - the amount which is proposed to be paid over by Canada on the transfer of the company's rights." This would indicate that Canada had no funds (at least, they weren't billions of dollars in debt in those days). The second item of concern in the negotiations was the aboriginals. "...It will be our duty to make adequate provision for the protection of the Indian tribes whose interests and well-being are involved in the transfer, and we authorize and empower the Governor-in-Council to arrange any details that may be necessary to carry out the terms and conditions in the above agreement." How many other promises have been broken since then?

The actual Deed of Surrender contained fourteen articles. The first starts off with the £300,000 sterling payment to the Bay. The second provides for the company retaining "all the posts or stations now actually possessed and may select blocks of land adjoining each of their posts or stations". This showed up in a lengthy schedule. Insofar as the Province of Alberta is concerned, then known as Saskatchewan Territory, Edmonton House[2] was granted 3,000 acres; Lac la Biche - 1,000; St. Albert -1,000 - for a total of 25,700 acres in the District. Strangely enough, only 1,500 acres were assigned to what we now know as the Northwest Territories, the largest being Fort Vermilion - 500 acres, Fort Nelson - 200 acres.

Of greatest interest to this study is the definition of the fertile belt bounded as follows: "On the south by the United States, on the west by the Rocky Mountains, on the north by the northern branch of the Saskatchewan River (North Saskatchewan R.), on the east by Lake Winnipeg, the Lake of the Woods, and the waters connecting them." It was in this expanse of millions of acres that Hudson's Bay

1 Queen Victoria.

2 H. H. Somerville states that when the Hudson Bay block in Edmonton was subdivided, the covenant prohibited anything except one-family houses. This legal obstruction was removed recently to allow condominiums and high rises.

Company was granted land "not exceeding one-twentieth part of the land so set out". Thus, from this arrangement, flowed the Section 8 and ¾ of Sec. 26 except for those townships divisible by five which would provide for the full Sec. 26 to make up the 1.8 sections (one-twentieth of 36 sections).

The agreement (Deed of Surrender) was signed November 19, 1869.

Appendix XVII

REDWATER HOMESTEADERS ON CROWN LAND

1894-1905
(Source: Lester & Anne Key)
(Verified by H. H. Somerville)

Year	Name	Map No.	Section	Twp	RgeW4M
1894	Barette, Ismail	1	SE26	56	22
1897	Jones, William	2	SW30	56	22
1899	Cook, Samuel	3	SE,SW,NE36	56	21
1899	Cook, Charles	4	NW36	56	21
1900	Andrews, Pierre	5	SW30	56	21
1900	Sawka, Alexander	6	NE34	56	21
1900	LePage, Enoch	7	NW30	56	21
1900	Huot, Onesime	8	SW19	56	21
1901	Lakusta, Fedro	9	NW35	56	21
1901	Picz, Jan	10	NW28	56	21
1901	Lakusta, John	11	SW2	57	21
1901	Ward, Peter	12	E/2 NE36	56	22
1901	Perraux, Marie	13	NW36	56	22
1902	Romaniuk, Ilja	14	SW35	56	21
1902	Romaniuk, Wasyl	15	NW30	57	21
1902	Piche, Nikolis	16	NE28	56	21
1902	Bridwaux, Joseph	17	SE30	56	21
1902	Scholes, James H.	18	SW28	56	22
1903	Staples, Moses	19	SE34	56	21
1903	Hunka, Mytro	20	SW34	56	21
1903	Pich, Dmytro	21	NE4	57	21
1903	Sawka, Rozalio	22	SE4	57	21
1903	Romaniuk, Nykolay	23	SE3	57	21
1903	Libbey, Jacob	24	NW2	57	21
1903	Hanlan, Sam	25	SE1	57	21

Year	Name	Map No.	Section	Twp	RgeW4M
1903	Stecyk, Matuij	26	SE22	57	20
1903	Derouin, Frank	27	SW26	56	22
1903	Shaw, John	28	SE28	56	22
1904	Puskaluk, Stephan	29	NE35	56	21
1904	Hunka, John	30	SW28	56	21
1904	Machowecki, Mike	31	NE26	57	22
1904	Kostiuk, Dmytro	32	SW26	57	22
1904	Kostiuk, Hryko		SW26	57	22
1904	Chmelnizhy, Wasyl	33	SE22	57	22
1904	Chmelnizhy, Dmytro	34	SW22	57	22
1904	Chmelnizhy, Zachary	35	NW22	57	22
1904	Wintoniak, Michael	36	NE22	57	22
1904	Osadczuk, Stefan	37	SE2	58	22
1904	Melnyk, Jurko	38	SW2	58	22
1904	Litwin, Stefan	39	NE26	57	20
1904	Laczowski, Iwan	40	NW26	57	20
1904	Halun, Michal	41	SW26	57	20
1904	Litwin, Alexander	42	SW22	57	20
1904	Moreav, E. F.	43	NW26	56	22
1904	Wilflingsider, Johan	44	NE34	57	22
1904	Antoniko, Dmitro	45	NE34	57	20
1904	Harmata, Ilan	46	SE34	57	20
1904	Litwin, Michael	47	NE22	57	20
1904	Litwin, Senko	48	NW22	57	20
1904	Czornyj, Iwan	49	SE24	58	20
1904	Paskaluk, Stephan	50	NW34	56	21
1905	Krajczyk, Joseph	51	SE18	57	21
1905	Cook, Maurice	52	SE30	57	21
1905	Hinkley, Clifton	53	NW20	57	21
1905	Ward, William	54	SE20	57	21
1905	Malowany, Petro	55	NE18	57	21
1905	Sywolos, Wasyl	56	NW12	57	22
1905	Kraychy, Wasil	57	NE12	57	22
1905	Lakusta, John	58	NE16	57	21
1905	Ward, Randolph	59	SW28	57	21
1905	Hanlan, Mitchell	60	NE32	57	21
1905	Fraser, William	61	SE6	58	21
1905	Cook, S. E.	62	SW6	58	21

Year	Name	Map No.	Section	Twp	RgeW4M
1905	Gibbs, E. W.	63	SW12	58	22
1905	Antasko, Michail	64	NW34	57	20
1905	Czorny, Fedko	65	SW34	57	20
1905	Moreau, Randolphe	66	NE30	56	21
1905	Ikacsyk, Mike	67	NW27	56	21
1905	Gagnon, Joseph	68	NW34	56	22
1905	Todd, R. H.	69	NW16	57	22
1905	Griffus, John Wm.	70	SE10	58	22
1905	Griffus, H. R.	71	SW10	58	22
1905	Schell, E. L.	72	NW10	58	22

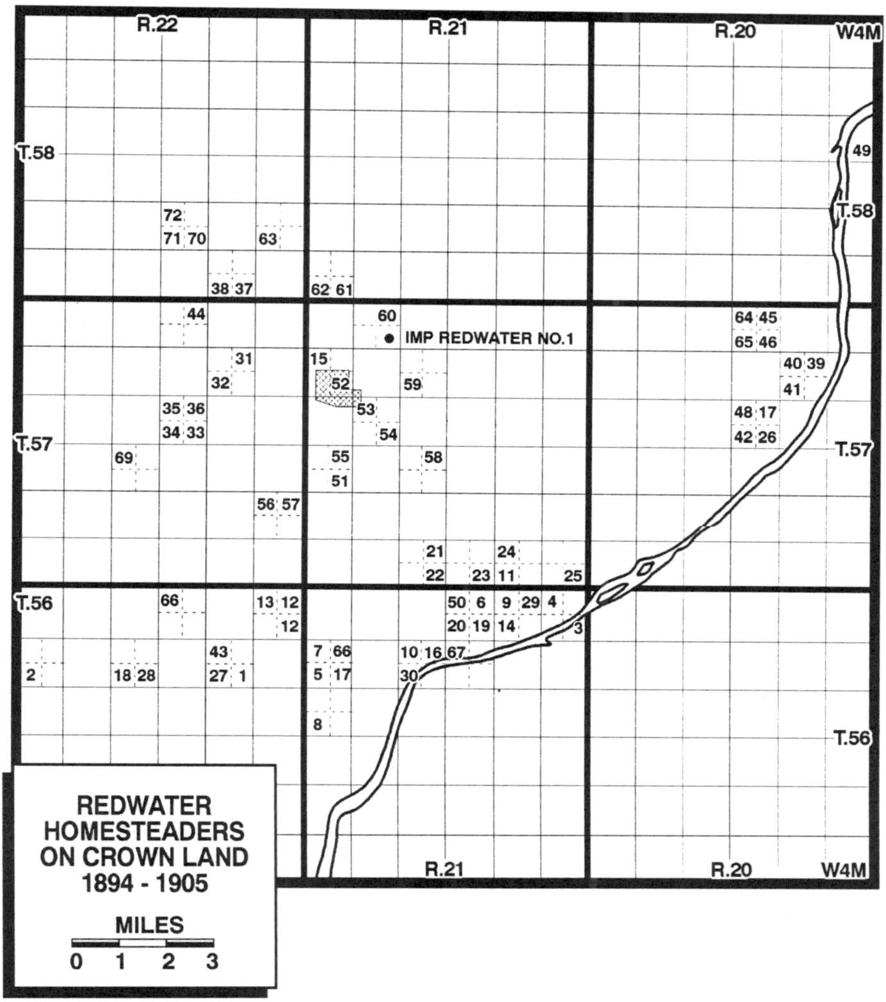

Appendix XVIII

REDWATER CROWN RESERVES SALES

1948 -1952
Courtesy H. H. Somerville

REDWATER
AREA

SALES OF CROWN RESERVES

PETROLEUM AND NATURAL GAS LEASES

Parcel	Acreage	Date of Sale	Name	Purchase Price	PER ACRE
1 - N.E.¼ 12-57-22-4	160	10/11/48	Rejected	$	
2 - N.E.¼ 24-57-22-4	160	10/11/48	Rejected		
3 - S.W.¼ 24-57-22-4	160	10/11/48	Rejected		
4 - N.E.¼ 34-57-22-4	160	10/11/48	Pacific Petroleums Limited (N.P.L.)	101,003.00	631.27
5 - N.E.¼ 36-57-22-4	159	10/11/48	Anglo-Canadian Oil Company Limited	351,025.00	2,207.70
6 - L.S. 3-36-57-22-4	40	10/11/48	Rejected		
7 - L.S. 4-36-57-22-4	40	10/11/48	Rejected		
8 - L.S. 5-36-57-22-4	40	10/11/48	Rejected		
9 - L.S. 6-36-57-22-4	40	10/11/48	Rejected		
10 - N.W.¼ 24-57-22-4	160	10/11/48	Rejected		
11 - S.E.¼ 24-57-22-4	160	10/11/48	Rejected		
12 - N.W.¼ 36-57-22-4	160	10/11/48	Pacific Petroleums Limited (N.P.L.)	283,003.76	1,768.77
13 - S.E.¼ 36-57-22-4	160	10/11/48	Home Oil Company Limited	302,039.00	1,887.74
14 - N.E.¼ 24-57-22-4	160	17/12/48	Pacific Petroleums Limited (N.P.L.)	127,077.77	794.24
15 - S.W.¼ 36-57-22-4	160	17/12/48	The British American Oil Company Limited	450,212.00	2,813.83
16 - N.E.¼ 15-57-21-4	160	15/2/49	Royalite Oil Company Limited (parcels 16–19)	3,153,491.00	4,927.33
17 - N.W.¼ 15-57-21-4	160	15/2/49			
18 - S.E.¼ 15-57-21-4	160	15/2/49			
19 - S.W.¼ 15-57-21-4	160	15/2/49			
20 - N.E.¼ 30-57-21-4	160	15/2/49	The British American Oil Company Limited and Hudson's Bay Oil and Gas Company Limited (parcels 20–23)	2,044,580.00	4,277.36
21 - N.W.¼ 30-57-21-4	160	15/2/49			
22 - S.E.¼ 30-57-21-4	159	15/2/49			
23 - S.W.¼ 30-57-21-4	159	15/2/49			
24 - N.E.¼ 8-57-21-4	160	29/4/49	Rejected		
25 - N.E.¼ 3-58-22-4	160	29/4/49	Pacific Petroleums Limited (N.P.L.)	91,099.00	569.37
26 - N.W.¼ 3-58-22-4	160	29/4/49	Rejected		
27 - S.E.¼ 3-58-22-4	160	29/4/49	Rejected		

- 2 -

REDWATER AREA

SALES OF CROWN RESERVES

PETROLEUM AND NATURAL GAS LEASES

Parcel	Acreage	Date of Sale	Name	Purchase Price	PER ACRE
28 - S.W.¼ 3-58-22-4	160	29/4/49	Rejected		
29 - S.E.¼ 10-58-22-4	160	29/4/49	Pacific Petroleums Limited (N.P.L.)	$ 201,099.00	1,256.87
30 - S.W.¼ 10-58-22-4	160	29/4/49	Pacific Petroleums Limited (N.P.L.)	61,099.00	381.87
31 - N.E.¼ 16-57-21-4	160				
32 - N.W.¼ 16-57-21-4	160	15/6/49	Amerada Petroleum Corporation	3,223,320.00	5,036.44
33 - S.E.¼ 16-57-21-4	160				
34 - S.W.¼ 16-57-21-4	160				
35 - N.E.¼ 8-57-21-4	160	15/7/49	G. E. Watt	263,130.00	1,644.56
36 - N.E.¼ 5-58-21-4	160	15/7/49	Texaco Exploration Company	396,000.00	2,475.00
37 - N.W.¼ 5-58-21-4	160	15/7/49	Texaco Exploration Company	587,000.00	3,668.75
38 - N.E.¼ 8-58-21-4	160	15/7/49	Rejected		
39 - N.W.¼ 8-58-21-4	160	15/7/49	Rejected		
40 - S.E.¼ 8-58-21-4	160	15/7/49	Rejected		
41 - S.W.¼ 8-58-21-4	160	15/7/49	Texaco Exploration Company	476,000.00	2,975.00
42 - N.E.¼ 10-58-22-4	159	15/7/49	Anglo-Canadian Oil Company Limited	626,000.00	3,937.11
43 - N.W.¼ 10-58-22-4	159	15/7/49	Texaco Exploration Company	413,000.00	2,597.48
44 - N.E.¼ 17-58-22-4	159	15/7/49	Pacific Petroleums Limited (N.P.L.)	97,698.00	614.45
45 - N.W.¼ 17-58-22-4	159	15/7/49	Pacific Petroleums Limited (N.P.L.)	92,698.00	583.01
46 - S.E.¼ 17-58-22-4	159	15/7/49	Rejected		
47 - S.W.¼ 17-58-22-4	159	15/7/49	Rejected		
48 - S.E.¼ 5-58-21-4	160	25/8/49	Royalite Oil Company Limited	650,121.00	4,063.26
49 - S.W.¼ 5-58-21-4	160	25/8/49	The British American Oil Company Limited	902,800.00	5,642.50
50 - N.W.¼ 3-58-22-4	160	25/8/49	Rejected		
51 - S.E.¼ 3-58-22-4	160	25/8/49	Ed-Bridge Petroleum Co. Ltd.	25,150.00	157.19
52 - N.E.¼ 9-58-22-4	160	25/8/49	Rejected		
53 - N.E.¼ 15-58-22-4	159	25/8/49	Texaco Exploration Company	501,000.00	3,150.94

- 3 -

REDWATER AREA

SALES OF CROWN RESERVES

PETROLEUM AND NATURAL GAS LEASES

Parcel	Acreage	Date of Sale	Name	Purchase Price	PER ACRE
54 - N.W.¼ 15-58-22-4	159	25/8/49	Texaco Exploration Company	$ 401,000.00	2,522.01
55 - S.E.¼ 16-58-22-4	160	25/8/49	Rejected		
56 - S.W.¼ 16-58-22-4	160	25/8/49	Rejected		
57 - S.E.¼ 23-57-21-4	160	25/8/49	Calvan Petroleums Limited	76,101.99 + 15%	475.64
58 - S.W ¼ 23-57-21-4	160	25/8/49	Calvan Petroleums Limited	176,105.99 + 15%	1100.66
59 - N.W.¼ 8-58-21-4	160	25/8/49	Alberta Giant Oil Company Ltd.	33,080.00 + 15%	206.75
60 - S.E.¼ 8-58-21-4	160	25/8/49	Rejected		
61 - S.E.¼ 17-58-22-4	159	25/8/49	The British American Oil Company Limited	100,160.00 + 15%	629.94
62 - N.E.¼ 12-57-21-4	159	27/10/49	Rejected		
63 - S.E.¼ 12-57-21-4	160	27/10/49	The Ohio Oil Company	420,031.00	2,625.19
64 - S.E.¼ 15-58-22-4	159	27/10/49	Tanner & Co. Limited	918,662.67	5,777.75
65 - S.W.¼ 15-58-22-4	159	27/10/49	Tanner & Co. Limited	677,267.17	4,259.54
66 - N.W.¼ 16-58-22-4	160	27/10/49	Canadian Gulf Oil Company	350,000.00	2,187.50
67 - S.W.¼ 16-58-22-4	160	27/10/49	Rejected		
68 - S.E.¼ 23-58-22-4	159	27/10/49	Rejected		
69 - S.W.¼ 23-58-22-4	160	27/10/49	Pacific Petroleums Limited (N.P.L.)	325,001.00	2,031.26
70 - S.E.¼ 19-58-22-4	160	27/10/49	Pacific Petroleums Limited (N.P.L.)	101,001.00 + 15%	631.26
71 - N.E.¼ 9-58-22-4	160	16/2/50	Polaris Oil Company Limited	125,117.27 + 15%	781.98
72 - N.E.¼ 19-58-22-4	160	16/2/50	Rejected		
73 - N.W.¼ 22-56-21-4	160	16/2/50	Atlantic Oil Company Limited	607,109.99	3,794.44
74 - S.E.¼ 22-56-21-4	160	16/2/50	Pacific Petroleums Ltd. (N.P.L.)	327,109.99	2,044.44
75 - S.E.¼ 34-56-21-4	160	16/2/50	Selburn Oil Limited	1,009,009.99	6,306.31
76 - N.W.¼ 36-56-21-4	160	16/2/50	Royalite Oil Company Limited	1,414,530.00	8,840.81
77 - S.W.¼ 36-56-21-4	160	16/2/50	Royalite Oil Company Limited	1,519,101.00	9,494.38
78 - N.E.¼ 1-57-21-4	160	16/2/50	Texaco Exploration Company	801,000.00	5,006.25
79 - S.E.¼ 1-57-21-4	160	16/2/50	Royalite Oil Company Limited	1,108,800.00	6,930.00
80 - N.E.¼ 16-58-22-4	160	16/2/50	C. L. McMahon, Inc.	1,011,701.53	6,323.13
81 - S.E.¼ 16-58-22-4	160	16/2/50	Texaco Exploration Company	521,000.00	3,256.25
82 - S.W.¼ 16-58-22-4	160	16/2/50	Bishop Oil Company	347,113.47	2,169.46

- 4 -

REDWATER
AREA

SALES OF CROWN RESERVES

PETROLEUM AND NATURAL GAS LEASES

Parcel	Acreage	Date of Sale	Name	Purchase Price	PER ACRE
83 - N.E.¼ 22-56-21-4	160	11/5/50	Pacific Petroleums Ltd. (N.P.L.)	$ 883,009.99	5,518.81
84 - S.W.¼ 22-56-21-4	160	11/5/50	Pacific Petroleums Ltd. (N.P.L.)	453,009.99	2,831.31
85 - N.E.¼ 34-56-21-4	160	11/5/50	Dome Exploration (Western) Limited	1,065,098.00	6,656.86
86 - N.E.¼ 36-56-21-4	160	11/5/50	Crown Trust Company	1,587,288.29	9,920.55
87 - S.E.¼ 36-56-21-4	160	11/5/50	Crown Trust Company	1,611,711.71	10,073.20
88 - S.E.¼ 8-57-21-4	160	11/5/50	Selburn Oil Co. Ltd.	477,777.77	2,986.11
89 - L.S. 13 of Sec. 26-56-21-4	40	11/5/50	Pacific Petroleums Ltd. (N.P.L.)	283,009.99 + 15%	7075.25
90 - N.W.¼ 31-56-20-4	160	29/6/50	Dome Exploration (Western) Limited	1,237,077.00	7,731.73
91 - S.E.¼ 31-56-20-4	160	29/6/50	Dome Exploration (Western) Limited	1,137,077.00	7,106.73
92 - S.E.¼ 33-56-21-4	160	29/6/50	Rejected		
93 - N.W.¼ 34-56-21-4	160	29/6/50	Pacific Petroleums Ltd. (N.P.L.)	926,101.01	5,788.13
94 - N.E.¼ 35-56-21-4	160	29/6/50	Pacific Petroleums Ltd. (N.P.L.)	1,827,101.01	11,419.38
95 - S.E.¼ 35-56-21-4	160	29/6/50	Crown Trust Company	1,853,333.00	11,583.33
96 - N.W.¼ 4-57-21-4	160	29/6/50	Pacific Petroleums Ltd. (N.P.L.)	556,101.01	3,475.63
97 - N.E.¼ 12-57-21-4	159	29/6/50	Rejected		
98 - N.W.¼ 18-56-20-4	159	24/8/50	Texaco Exploration Company	502,000.00	3,157.23
99 - S.W.¼ 29-56-20-4	160	24/8/50	Dome Exploration (Western) Limited	633,174.00	3,957.34
100 - N.E.¼ 31-56-20-4	160	24/8/50	Dome Exploration (Western) Limited	927,174.00	5,794.84
101 - S.W.¼ 31-56-20-4	161	24/8/50	Royalite Oil Company Limited	1,727,100.00	10,727.33
102 - S.W.¼ 34-56-21-4	160	24/8/50	Pacific Petroleums Limited	500,101.99	3,125.64
103 - N.W.¼ 35-56-21-4	160	24/8/50	Royalite Oil Company Limited	1,423,800.00	8,898.75
104 - S.W.¼ 35-56-21-4	160	24/8/50	Royalite Oil Company Limited	1,628,100.00	10,175.63
105 - S.W.¼ 4-58-21-4	159	24/8/50	Royalite Oil Company Limited	225,900.00	1,420.75
106 - S.W.¼ 6-57-20-4	160	2/11/50	Dome Exploration (Western) Limited	937,000.00	5,856.25
107 - N.E.¼ 10-56-21-4	160	2/11/50	Canadian Gulf Oil Company	75,000.00	468.75
108 - N.W.¼ 27-56-21-4	160	2/11/50	Scurry Oils, Limited	276,830.00	1,730.19

- 5 -

REDWATER AREA

SALES OF CROWN RESERVES

PETROLEUM AND NATURAL GAS LEASES

Parcel	Acreage	Date of Sale	Name	Purchase Price	PER ACRE
109 - N.E.¼ 33-56-21-4	160	2/11/50	Scurry Oils, Limited	$ 302,830.00	1,892.69
110 - N.W.¼ 8-57-21-4	160	2/11/50	Rejected		
111 - N.W.¼ 23-57-21-4	160	2/11/50	Pacific Petroleums Ltd.	250,101.01	1,563 13
112 - S.W.¼ 19-58-22-4	160	2/11/50	J. B. Barron and A.L. Barron	42,110.00	263. 19
113 - N.E.¼ 18-56-20-4	157	10/5/51	Scurry Oils Limited	80,833.00	514.86
114 - S.W.¼ 18-56-20-4	160	10/5/51	Superior Oils Limited	125,111.00	781.94
115 - N.W.¼ 6-57-20-4	160	10/5/51	Dome Exploration (Western) Limited	569,009.00	3,556.31
116 - S.W.¼ 7-57-20-4	160	10/5/51	Superior Oils Limited	65,111.00	406 94
117 - N.W.¼ 10-56-21-4	160	10/5/51	The Bay Petroleum Corporation	227,345.15	1,420.91
118 - S.W.¼ 4-57-21-4	160	10/5/51	Hales H. Ross	225,478.60	1,409.24
119 - N.W.¼ 8-57-21-4	160	10/5/51	Dalhousie Oil Company Limited	52,530.00	328.31
120 - N.E.¼ 12-57-21-4	159	10/5/51	Scurry Oils Limited	185,831.00	1,161.44
121 - S.E.¼ 24-58-23-4	160	10/5/51	Scurry Oils Limited	50,830.00	317.69
122 - N.W.¼ 9-56-21-4	160	8/11/51	Admiral Oils Limited, Renfrew Petroleums Limited and Trans Empire Oils Limited	241,200.00	1,507 50
123 - N.E.¼ 11-56-21-4	160	8/11/51	The Bay Petroleum Corporation	217,720.20	1360.75
124 - N.W.¼ 11-56-21-4	160	8/11/51	The Bay Petroleum Corporation	227,130.30	1,419.56
125 - N.W.¼ 12-56-21-4	160	8/11/51	The Bay Petroleum Corporation	87,666.66	547.92
126 - N.E.¼ 5-57-21-4	160	8/11/51	Ross and Sons Oil Limited	53,725.80	335.79
127 - N.W.¼ 9-58-22-4	160	8/11/51	Ross and Sons Oil Limited	43,940.00	274.63
128 - S.E.¼ 9-58-22-4	160	8/11/51	Deep Rock Oil Corporation	35,000.00	218.75
129 - N.E.¼ 19-58-22-4	160	8/11/51	Albercan Oil Corporation	331,160.67	2069.75
130 - S.E.¼ 30-58-22-4	160	8/11/51	Oilans Petroleums Limited (N.P.L.)	36,210.00	226 31
131 - N.W.¼ 20-56-20-4	163	22/5/52	Rejected		
132 - S.W.¼ 20-56-20-4	163	22/5/52	Rejected		
133 - N.W.¼ 29-56-20-4	160	22/5/52	Rejected		

- 6 -

R E D W A T E R
A R E A

SALES OF CROWN RESERVES

PETROLEUM AND NATURAL GAS LEASES

Párcel	Acreage	Date of Sale	Name	Purchase Price	PER ACRE
134 - N.W.$\frac{1}{4}$ 32-56-20-4	160	22/5/52	Rejected		
135 - S.W.$\frac{1}{4}$ 32-56-20-4	160	22/5/52	Rejected		
136 - S.E.$\frac{1}{4}$ 33-56-21-4	160	22/5/52	Scurry Oils Limited	$ 72,000.00	450 00

Appendix XIX

Ukrainian Beliefs

(based on conversations with the Malowany family)

by the Author

The history and traditions of the Ukrainian people have always been inextricably linked to their religious beliefs.[1] Poland had routinely looked to Rome for her faith and this allegiance to the Pope reflected on the neighbouring Ukrainians, who had embraced Byzantine, or Eastern rite liturgy. This split was exacerbated by Rome's insistence on Latin while the Greek, or Byzantine, were permitted to use the vernacular. But political and ethnic tensions contributed to their failure to understand one another. When Poland seized a part of Ukrainian homeland, she further intensified her efforts to Latinize. One could say that the Ukrainians were trying to enjoy the best of both worlds, moving to Roman Catholicism - although controlled by the Polish landlords - but retaining the liturgy of the Orthodox. They wanted to retain "Church Slavonic," an otherwise dead language. In 1595 the Bishop of Lvov petitioned the king of Poland that his flock wished the Pope of Rome to be their head rather than the patriarch of Constantinople. Out of this emerged the 1596 Treaty of Brest-Litovsk. It was an attempt to address the complex socio-economic problems facing the Ukrainians while retaining the spirit of their eastern rites. An Orthodox bishop was appointed to the metropolitan See in Galicia and the Orthodox church was able to reestablish some of its authority. This new arrangement gave the Ukrainians the same privileges as Roman-Catholics.

Looking back over their troubled history it is remarkable that Christian religious beliefs, whatever the nuances, withstood repeated batterings. These started when Mongol hordes raped, pillaged and destroyed Kiev. Liberties were abrogated in the eighteenth and nineteenth centuries because of the partitioning of Poland. The situation under Russian rule following World War I was even worse than that under Austria-Hungary's yoke, nothing more than a religious vacuum. In 1946, the Russians abolished the Ukrainian Church, declared it illegal, and stated that it did not exist. The banning

---

1 See "A Brief Historical Note" by Peter Yachimec, Toronto, 1976, which follows.

and prohibition of books written in Ukrainian was an inevitable and devastating outcome. Other beliefs were punished as well, and the only faith tolerated was Russian Orthodox. More recently Gorbachev wanted all church property to be divided up, a matter still being fought over.

Serfdom was a way of life at the turn of the century, and this oppression, as well as the requirement to serve in the Austro-Hungarian Empire's army, persuaded many young men to skip to Germany and sail to the New World. Extreme inequality existed between the landlords and serfs. The serf paid tithes and taxes on the land and could neither sell it nor divide it among his heirs without the lord's permission. He was even forced to buy a specific quantity of whiskey from the lord's distillery. Peter Yarmula may have been one of those who lived under this repression, and maybe that's why he was so expert at making his own white lightning on his farm outside of Redwater.

A brief historical note

The village of Staryawa in Ukraine where the Yachimecs and the Bushkos were born is located in a western province of Ukraine near the border of Poland. This province was called Halychyna and its capital city is Lviv or Lvov.

Halychyna borders with White Russia in the north, Poland on the north west, Czechoslovakia and Hungary on the west, Romania and Moldavia on the south.

For centuries Ukraine had unstable governments and was an easy prey to invading outsiders wanting their rich agricultural land. The country was repeatedly overrun by Poles, Scandinavians, Russians, Turks and Tartars.

Christianity came to Ukraine from the Greeks through Byzantium(Constantinople). The first Christian church in the neighbouring countries was also Greek Orthodox, however, Rome succeeded in converting these people into Roman Catholicism, working their way through Germany, Austria, Hungary and the other neighbours. In order to convert the Christians in Ukraine Rome resorted to various schemes. First they sent missionaries which had little success. They then plotted with the rulers of Hungary and Poland to achieve their ends. Poland waged wars of aggression over the area of Halychyna with the Pope's blessing and support. This resulted in control by Poland over extended periods.

During these periods of occupation pressure was exerted on the people of Halychyna to become Roman Catholics and to use the Polish language. Polish settlers were brought into many villages. Roman Catholic churches were built. Various inducements were employed to win over as many as possible to Roman Catholicism. A number of Ukrainians did change over as time went on, taking advantage of bribes of land, better treatment and civil service jobs.

Late in the 16th century there began a movement among some of the Ukrainian Greek Orthodox Bishops to consider a formal union with Rome. This eventually resulted in the "Beresteyska Unia" in 1594 which formally turned the churches in Halychyna to Rome.

There was considerable opposition to this by the ordinary people so a compromise was agreed to by Rome giving the churches in Ukraine the right to continue with their existing customs, that is, the use of their language in church services, their traditional ceremonies and married priests.

The Pope in Rome was now in full control. The Latin language was used for administrative purposes, records and correspondence. This explains the use of Latin on such documents as birth certificates. Also since Poland had political control over Halychyna the Polish language often appears on official documents as well.

Usually Ukrainian names were given Latin versions by someone who thought he knew what they should be in Latin. Thus the name Pelagia or Palagia was the Latin for Pazka or Pazia. Also Theodori was for Fedir or Fedor or Teodor.

Ukrainians who emigrated to other countries continue to have these problems as other languages are encountered. Fortunately they make an effort to preserve their original language and the Cyrillic script.

Toronto 1976 Peter Yachimec

Appendix XX

"MEMORIES"
REDWATER & DISTRICT

Book Review by the Author

Printed 1972 by Friesens, clothbound, 227 pp
Second printing, 1972; Third printing, 1980.
Press runs not known - out of print.

There is no mention of Alberta Government help in the Acknowledgements. Two donations are cited: $100 from the Town of Redwater, $332 from the Redwater Centennial Committee.

"Errors have been made ... since no one is infallible, we hope the errors which you may find will be forgiven."

Foreword:

We sincerely hope it is worthy of all who helped in their own way ... It has been our aim to pay tribute to them and to present an account of their lives. In the "Builders of the Book," there is no mention of Anne Walker (nee Key) although articles by Anne appear in the body of the volume.

Early History (Page 1)

"Surveyors started in 1882, setting aside Hudson's Bay and School Lands. Survey completed in 1906, and up to that time, land was not available for homesteading." (List of homesteaders [see Appendix XVII] refutes this and the following quote of the first settler in 1894.)

> On Sept 7, 1906, land in the Redwater area was available for filing. A $10 fee was required One had to be a male, 18 years or over, who was or intended to become a British subject - or a widow with dependent children. A habitable dwelling had to be built, and the homesteader had to live on his property for at least three months of every year for three years. Thirty acres of land were to be cleared and broken, and fences had to be built to keep out wandering cattle and horses. Some of the settlers who arrived before 1906 simply squatted on the land they liked and filed when the survey was completed.

Why did the French settlers come from Rhode Island? Could they have been Quebecois who had moved to the States to be labourers in New England factories?

The Victoria Trail: One of the principal means by which the settlers travelled from the Redwater Area to Edmonton. To quote the book, "Winding trails following high ground over which many tired feet had trudged - they are gradually being erased - but dear God, don't let our heritage disappear!

Redwater had its beginning in 1918, when a flag stop was created on the newly constructed CN rail line. The name Redwater is attributed to a Mrs. Hannah Brown, who is said to have first noticed the colour of the water in the stream, apparently from iron oxide.

The settlers weren't the only people scouting the country. Banks sent their zealous, young juniors out to beat their competition to the punch by setting up branches in tents and many other structures. Let an unidentified bank employee with the Bank of Commerce relate his experiences.

> "About noon we successfully located the townsite of Redwater, with one general store, a log building about 12 ft x 20 ft, and made arrangements with the Russian proprietor for temporary premises until our tent, equipment and furniture arrived. Mr. Anderson obtained board and lodgings at a farm house nearby, and after bidding him a fond farewell, Mr. Nelson and myself proceeded to penetrate further into the wilderness. About 3 p.m. we located the townsite of Waskatenau, where we found that the same number of building permits had been issued as at Redwater, but the general store here was a farm building about 20 ft x 30 ft and was already decorated with a canvas sign reading, 'The Standard Bank of Canada'.
>
> "The following Tuesday ... Mr. Nelson and I started out again by taxi to locate Radway. We had proceeded about forty miles, when our driver tore the rear end out of his car, while driving through a sandy stretch of road. ... We eventually arrived back in Edmonton at 6 p.m.
>
> "The following morning the three clerks who were to act as our bodyguards were at the C.N. yards bright and early, and the train eventually pulled out about 11 a.m. Gibbons was our first stop, where ... we were to pitch a tent ... and we had a portion of our equipment thrown out to the side of the track.... we proceeded to locate a good brush patch which would act as a windbreak for our tent. ... Our first duty was to have lumber hauled from Bon Accord for the floor and office fixtures, and as there was no carpenter in the district it was necessary for us to do our own work. Our labour would hardly have

passed inspection by the Supervisor of Bank Premises, but it served the purpose, temporarily, at a minimum cost."

Quoting further. . .

"When in 1949, Devon Estates started to build the first of its 109 homes on land purchased from Mr. N. Ziniuk in the M.D. of Smoky Lake, the company built a water line from the Saskatchewan River to the housing development. It wasn't, however, until 1955 that the rest of the town began to enjoy the convenience of running water. That year Devon Estates sold the water system to the town for one dollar. Water and sewer extensions were then made to both sides of the town. With the number of water users increasing rapidly, a Horton watersphere reservoir was built. It had a capacity of 50,000 gallons and was the first of its kind in western Canada built by that firm.

"Over the years as the population of the town grew sizeably, and as the concern expressed about the potability of water drawn down stream from Edmonton became critical, a water pipe line from Edmonton seemed to be the only answer. It veered its way to Fort Saskatchewan and then to Redwater and came into operation in November 1970. It insured, even in the hottest summer, a steady supply of good water for the residents of the town of Redwater, a town that oil put on the map.

"Whatever the original residents may have said about the thunder of big trucks and the groan of the oil rigs, and the influx of strangers and wild roughnecks, and the always present and ever noisome petroleum gas, the sight and smell and feel and taste of money brought in by the new industry was sweet to all.

"Men fevered in the agony of searching for oil and then in the agony of finding of it. The grating roar of big diesels probing in the bowls (sic) of the earth was unceasing. Oil derricks lit up the night sky like a dozen midways. The whole scene had about it the air of a dream, the air of the unreality of the Big Tent. And in this seething circus of human affairs there was the comedy, no less than the tragedy; the ecstasy, no less than the agony. With the ecstasy that humans find in growing things there is also the agony. Redwater, before the oil boom, had a population of about 160, including the few farm families living on the periphery of the town."

In 1953 production was 80,000 barrels a day, and in 1956 a gas plant was built that cost $3.5 million. This is the one that Imperial said would bankrupt them.

"The arrival of the angels of mercy in the name of The Sisters of Charity of the Immaculate Conception, a Canadian order, and a doctor to Radway heralded the phase of modern medicine to the early pioneers. They brought a particularly appealing tradition to Radway when they opened their only Alberta hospital in 1926. In 1926, three Sisters came to Radway to found the St. Joseph's Hospital which seemed so remote that it was referred to as a foreign mission. There was a small and very inadequate house where the first three Sisters literally "pitched camp" and set up to serve an urgent need as best they could. They set aside as much of the house as possible for patients, reserving for themselves only one room which was to serve as dormitory, community living room, and in times of pressure, a dining room. There was a lean-to which had been added to the main building for a chapel, and a room which they called Refectory; but when this refectory was in demand for surgery the Sisters reverted to their single many-purpose room and ate with their trays on their knees. For each major surgery, the refectory had to be thoroughly cleaned, walls wiped down, tables sterilized --a rather extensive "scrub" indeed.

"At 35 degrees below zero, the bedding still had to be washed. Pioneering in isolated districts was always difficult and the rates, when paid, ranged from 50¢ a day to $2.00 per day. Many were unable to pay their bills with cash, but the Sisters were very happy to receive vegetables, chickens and even the odd live hog, even though they did have their own garden, cows, hogs and chickens."

NOWERS INTERVIEW relates the story of his having offered the Sisters $3.00-plus per day for the Imperial Oil employees who received hospital care. They refused, saying that $2.00 was their going price.

BIBLIOGRAPHY

Andrichuk, John (1958), Stratigraphy etc. - Redwater etc., American Association of Petroleum Geologists, Vol 42, No. 1, pp 3–93.

Andrichuk, John (1961), Tectonics - Duhamel, American Association of Petroleum Geologists, Vol. 145, No. 5.

Baugh, J. E. (1950), "The Redwater Field," Canadian Oil & Gas Industry (May), pp 22–25.

Baugh, J. E. (1950), "Reservoir Data on the Redwater Field," World Oil, Int. Sec. (December), pp 260–264.

Baugh, J. E. (1951), "Leduc D-2 Zone Pool," World Oil, Int. Sec. (February), pp 210–215.

Baugh, J. E. (1951), "Leduc D-3 Zone Pool," World Oil, Int. Sec. (March), pp 210–214.

Beach, F. K., and Irwin, J. L. (undated, Pre-Leduc), "The History of Alberta Oil" issued by Dept. of Lands & Mines.

Corey, B. H., "Petroleum & Natural Gas, Summary Western Canada Sept 15–Oct 15 (1948), CIM Bulletin, pp 651–654.

Hanson, E. J. (1958), "Dynamic Decade, Chp #8 - Redwater," publ. by McClelland & Stewart, Ltd., Toronto.

Haskett, I. (1951), "Reservoir Analysis of Redwater Pool," The Oil & Gas Journal, Vol. 50 (October).

Klovan, J. E. (1966), "An Integrated Method of Facies Analysis and Reservoir Analysis as Applied to Redwater Field," A6, The Computer Sec. of Canada, Fifth Nat. Conf., Banff, AB.

Tixier, M. P. and Forsythe, R. L. (1951), "Application of Electrial Logging in Canda," CIM Vol. LIV, pp 358–368.

Visser, C. and Branscombe, A. (1951), "Drilling Symposium," COGI Techn Soc. Papers, pp 1–3, #1 - Turner Valley & Foothills Drilling (includes Redwater (Jan–Feb).

ACKNOWLEDGEMENTS

Alberta Legislature Library Staff
Alberta Provincial Archives Staff
Alert, Betty
Allen, Bill
Andrichuk, John
Archibald, Ralph G.
Armstrong, Jack
Atkinson, Ralph
Astle, Gordon & Rita

Bader, Karen
Baillie, A. D.
Ballachey, John
Barber, Rob
Barlow, C. B./C. B., Jr.
Barroll, A. E.
Barron, Jim
Bediz, Peter
Belot, Gordon
Bischoff, R. M. (Bob)
Bilandzija, Paul
Binder, Esther
Black, Frank
Blackadar, Murray
Bleakley, Steve B.
Boston, Sophie
Bowe, Wilf & Audrey
Bowsher, A. P. (Pat)
Browning, Jack
Bruce, John E.

Carpenter, George
Chaba, Alex & Nancy
Chaba, Walter & Theresa
Chapman, Carl
Chekerda, John
Chekerda, Larry
Chen, Ed
Chesher, Ian
Chobotuk, J. W. (Jim)
Christopher, Voula

CIM Calgary Office
Buchanan, Catherine
Fancey-Kelly, Marina
Magnuson, Kari
Munro, Gloria
Story, Sheila
Walton, Lynette

Clark, W. F. (Bill)
Cleveland, Courtney
Collins, Neal
Connell, Gordon
Cook, Hilton & Christina
Cook, Morris
Coote, G. F. (Bud)
Corey, B. H. (Bert)
Cormack, Gord
Cormack, Si & Muriel
Crawford, Ian
Court, Ted & Pearl

Dabbs, W.
Daily Oil Bulletin
Dranchuk, Peter
Duczek, R. L. (Dick)

Edgecombe, Rod
Edie, Ralph
Edwards, Garnet
Elliott, R.H. J. (Bob)
Erdman, Oscar
Evans, Paul C.
Evasiw, Jim & Anne

Findlay, R. J.
Forry, Eve
Foster, Glenn

Gallagher, Jack
Gamble, Doug
Gammell, H. G.

Gantz, Perla
Gavel, Elmer & Eleanor
Gibson, W. J. (Gibby)
(posthumously)
Glenbow Museum
Bryant, Ellen
Cass, Doug
Garnier, Andrea
Kooyman, Susan
Moir, Lindsay
Molesky, Pat
Myhr, Catherine
Ryckman, Roberta
Walton, Lynette
Greenwood, Paul

Hall, Don & Lois
Hancock, W. P. (Bill)
Harding, S.R.L. (Stan)
Harvie, Don
Haskett, Irene
Haugrud, Gunnar
Hawkins, Gordon
Hawrelko, John
Hay, Gladys
Hobson, George
Hochhausen, E.F.D. (Ben)
Hosegood, Bill
Hoyda, Don
Hrynchuk, John & Rose
Hrynchuk, Larry
Hrynchuk, Mike
Hrynchuk, Walter & Vera
Hunchak, John & Lucy
Hunter, Beatrice
Hunter, Don
Hunter, V. H. (posthumously)

Interprovincial Pipe Line Inc.
Irwin, Jim

Jaremko, Gordon
Jeffery, Bernard & Kay
Jorgensen, Alexandra

Key, Anne & Lester
Kidd, Fred
Kimura, Jim & Chizuko
Kinton, Dave
Kiyooka, George
Knapp, Hal
Knowles, R. D. (Doug)
Kooner, Kalvinder
Krull, Gerry

Laborde, Ed
Lakusta, Ed
Lakusta, Helen
Lambert, John
Lambert, Ed & Barbara
Langston, J. F. (Spi)
Lavoie, Doug
Leiper, Hugh
Leitch, George
Lineham, Fin
Loranger, Diane

Malowany, Mary
Manyluk, Frank
MacKenzie, Hugh
Mason, A.D.M. (Tony)
McArthur, Doreen
McCaskill, Jack & Kay
McCauley, Patrick
McClintock, Mrs. Marjorie
McCourt, G. B. (Garry)
McKinnon, Ian
McLaws, D. P.
McLaws, I. J.
Mech, Bonnie
McPhee, W.
Morris-MacRae, Mrs. Jean
Mugler, Bill
Murray, Roy
Mutch, J. P.
Myroon, John & Olga

Nesbitt, Jack (posthumously)
Nesbitt, John
Nowers, Lauder (posthumously)

Oakley, Ken
Owre, Slim

Patterson, Bruce
Priddle, Roland
Pasemco, Vic & Beanie
Paulson, Maurice
Payne, Alice
Perkins, Vicki
Pettinger, J. B.
Priddle, Roland

Rabey, Wes
Rault, Corrine
Roberts, Frank
Roloff, Rudy & Myrt
Rowland, Les (posthumously)
Royer, Yvonne & Rita
Rozak, Nick &Katherine

Sawka, Mary
Scott, Jim
Senych, Mike
Shaw, E. W. (posthumously)
Shaw, John, Jr.
Siferd, Clark
Sluzar, R. D. (Ray)(Sluz)
Smith, Melvin
Somerville, H. H. (Hubert)
Stelck, C. R. (Charlie)
Stetsko, Matt & Helen
Strand, Jim

Taylor, Nick, MLA
Taylor, John
Teske, Gord
Torguson, Ray & Norma
Treleaven, Marian L.
Tull, Ray & Lois

Unal, Jade

Walker, Carolyn
Wark, Tom
Warne, George
Watson, Evelene
Webster, Gordon
Webster, J. H. (Harry)
Wilkins, Kitty
Wright, Marge
Wyatt, Brian

INDEX

Abzinger, Julius 39, 89, 207
Addison, Sandy 48, 275, 276
Aird, Campbell 92,196
Allen, W.P. (Bill) 191, 201
Alton, Ken 183
Amelia (well - formerly Redwater #3) 31, 159, 168
Amerada -Hess 33, 136, 137, 168, 258, 289, 292
Anderson, Wilbur 17, 18, 200
Andrichuk, John 63–65, 67, 315, 373
Anglo-Home 35, 72, 91, 140, 161, 162, 229, 230, 270, 289, 291, 294
Archibald, Ralph 55
Armstrong, Jack 10, 30, 31, 188
Astle, Gordon 190, 218, 311
Astle, Rita (nee Cook) 15, 216–218
Atkinson, Bill 172
Atlantic No.3 (well) 3, 23, 49, 52, 163, 173, 243
Atlantic Richfield 260

Bailey, Alex 161
Bailey, Charlie 84
Baillie, A.D. (Andy) 69
Baillie, Wilf 152, 213
Bain, Art 90, 92
Baker, Don 51
Baker, Sheron 51
Banister, Chick 167
Bannantine, George 80, 81, 89, 121, 147, 207, 251
Barlow, C.B. ("Boss") 73, 145–147, 255
Barnsdall Honolulu Seaboard 98, 110, 137, 138, 140, 144, 172
Barroll, Alfred Edwin (Ed) 39
Barron, Jim 13
Baugh, J.E. (Ted) 85, 321, 322, 373
Beard, Gordon 15
Bear Oil 34, 85, 173
Beck, Gus 46
Bediz, Peter I. 10, 13, 18, 135, 154
Bevel, Jack 143
Bigoray, Mr. 42
Binning, Ralph 51
Bird Construction 119, 184
Bischoff, Robert M. (Bob) 110, 138, 140
Bishop, H. 140
Black, Frank 82–84
Blackwell, "Kayo" 37, 38, 53
Blakeman, Denis 192, 193
Bleakley, Steve 36, 121, 194, 195, 197, 198
Boggs, Oswald Desmond (Des) 9, 61
Bowe, J.W. (Wilf) 22, 83, 311

Branscombe, Art 50, 373
Brascan 142
British-American (B-A) 87, 90, 165, 171, 184, 220, 247, 258, 259
Brinkerhoff Drilling 186
British Empire Oil 47
British Petroleums (Triad) 171, 173, 230
Bronfmans 149, 165, 169
Brookshire, Mr. 152
Brown, Jr., R.A. (Bobby) 172, 173
Brown & Root 122
Browning, Jack 85, 152, 153, 173–175
Bruce, Johnny 83
Budney, Joe 140
Burn, Ivone St. G. 148, 149, 158, 169, 170, 207, 208

Cal Standard/Chevron 13, 61, 153, 213, 250
Callow, George O. 76
Calvan Consolidated 163, 171
Campbell, Bob 260
Campbell, Ed 147
Can-Tex Drilling 30, 37
Canada West Drilling 32
Canadian Fina Oil Ltd. 47, 48, 135, 142, 210, 227
CANOL Project 61, 111
Canuck Drilling 47

Cardwell (Rigs) 46, 47, 104
Carlyle, Harry 112
Carpenter, Jim 15
Carsted, Douglas J. 75, 76
Carter Oil 10, 87, 90
Century Geophysical 33
Chaba, Alex 187, 197, 199, 209, 210, 220
Chaba, Nancy 202, 209
Chaba, Peter 259
Chaba, Walter and Theresa 189, 210, 211
Chambers, Everett 167
Chapot, Ernie 51
Charlton, Tom 311
Chekerda, John 3, 194, 211–212, 281
Chekerda, Larry 4, 5, 211, 212, 281
Chesher, Cec 152–153, 161, 163, 213–215
Clare, Barney 14, 290
Clark, Alex 150
Clark, Les 66, 143–145, 150, 152, 153, 174
Clark, W.S. (Bill) 202
Cleveland, Courtney 143–145
Collins, H.N. (Neal) 84, 85, 250
Commonwealth Drilling 47, 52
Companies Act 94
Connell, Gordon (Gord) 35, 36, 88, 92, 98, 162, 165–167, 169
Conservation Board, Alberta - see Energy Resources Conservation Board (ERCB)
Constable, Leigh 80, 81, 107, 196
Cook, Amelia 31, 216, 217
Cook, Hilton 16, 57, 216–218
Cook, Jack 15, 216
Cook, Jud 15, 216, 218
Cook, Maurice 15, 216, 218
Cook, Rita - see Astle, Rita
Cook, Samuel 31, 216, 217
Coote, Bud 14, 30, 301
Corey, Bert 52, 373
Cormack, Gord 111–114
Cormack, Mildred 121
Cormack, Si 89, 111, 112, 196
Court, Pearl and Ted 184, 219–221
Cousins, Bill 89, 104, 111
Crawford, George 148, 154
Crawford, Ian 135, 152, 153
Crow, Pete 84
Curran, R.B. (Rapid Robert) 40, 79, 80, 119, 172, 173, 207

Dallas, Jack 130
Darling, Gordon 294
Darwin, Charles 61, 71
Davis, Fred 14
Davis, Warren 91, 95
Dawson, Keith 104, 105
Dawson, Rex 149
Deacon 195
De Koch, William 145, 146
Desjardins, Lanky 42, 43
Desnoyer, Al 21
Desrosier family 56
Devon Estates 141, 371
Dingle, Mabel 111
Dingle, W.B. (Walt) 111, 122
Dixon, R.K. (Bob) 51, 52
Doig, Ian 126
Dome 141
Donaldson, Lynn 47
Donnelly, Peggy 199
Dowell 21
Dowhaniuk, Diane 112–114
Downing, John 63, 72

Eason, T.T. 39, 40
Eason, T.W. 39
Edgecombe, Rod 80
Edie, Ralph 64, 65
Edmunson, Oly 104
Edwards, Garnet 52
Egremont (town) 31, 182, 183, 193, 212
Elliott, R.H.J. (Bob) 36, 37, 194
Empire Trust 137, 141, 147
Emslie, Roy 51
Energy Resources Conservation Board (ERCB) 23, 96, 122, 125, 130, 163
Environmental Protection and Enhancement Act 125

Erdman, Oscar 252
Evans, H.E.M. (Harry) 133, 134
Evans, Paul and Sarah-Beth 43–47, 143
Evasiw, Anne (nee Zeniuk) 200, 222
Evasiw, Jim 181, 187, 222
Evasiw, John 149

Falcon Seaboard Drilling 40
Falconer, Lorne 33, 173
Federated Petroleums 172
Fedorah (well) 13, 283, 289, 292, 301
Fischer-Tropsch 10, 14
Fisher, George 82
Flanders, Ralph 89, 184
Flewelling, Frank 49–51
Forsythe, R.L. (Russ) 85, 86, 373
Fournier, Frank L. 71, 72, 74, 85, 323
Fox, Glen 72
Fox, Lorraine 67, 185, 186
Fruman, Ernie 202
Fulmer, Ed 153

Gallagher, Eddie 44
Gallagher, Jack 81, 141, 142
Galloway, John O. 153
Gallup, Bill 165
Gamble, Doug 3, 17, 19, 20, 251, 311
Gamble, Glenn 21
Gavel, Elmer and Evelyn 55–57, 190
General Petroleums 48, 49, 172, 293
Geological Survey of Canada (GSC) 61, 62
Gibson, Kelly 51, 244
Gibson, Willis J. (Gibby) 15, 55, 56, 81, 87–89, 92, 97, 102, 118, 148, 183, 188, 192, 263, 264
Gilmore, R. (Bob) 90
Glendinning, Lou 50
Godwin, Jack 45, 104, 106
Goodall, D.P. (Red) 32
Goodman, A.J. (Jack) 62
Goodman, Nate 22
Grant, George 13
Green Hornet Report 92, 337, 338
Greenwood, Paul 95–97, 193
Grier, Bob 12
Griffith, Wilbur 45
Gulf Oil 42–44, 90, 91, 111, 112, 142, 143, 171, 244, 252, 281
Gunn, Warren 173, 247
Gustafson, J.D. (Dave) 22, 30, 32, 82, 90

Haffner, Ray 201
Haider, Mike 32, 79, 199, 207
Halliburton (Company) 21, 192, 247
Hally, Mrs. 202
Hancock, W.P. (Bill) and Doris 9, 14, 15
Harris, "Dollar for Dick" 30, 37
Hartling, Roy 147
Harvie, Donald S. 84, 133, 135, 136, 138, 141
Harvie, Dorothy Jean (nee Southam) 114
Harvie, Eric 35, 91, 119, 133–138, 140–143, 145, 147, 151, 162, 168, 283, 292
Harvie, Jack 30, 80, 97, 115, 116, 141, 183, 195
Harvie, D.M. (John) 161, 164, 167
Haskett, Irene 87, 88, 90, 95, 167, 173, 321, 325, 373
Haugrud, Gunnar and Rose 90–92, 194, 337
Hawke, Charlie 31, 32
Hawkins, Gordon 132, 134, 135, 148, 149, 158, 170, 201, 202
Hawkins, Jim 124
Hawrelko, John 179, 203, 224, 225, 235
Heard, S.F. (Bobby) 167
Heiland Exploration 35, 84, 152, 153, 213, 247, 268, 290, 301
Henderson, Allan 51
Henderson, Jim 89, 97, 197
Henkelman, Ida 14
Henley, Jack 88, 322
Hess, Gordon R. 34, 152, 158, 163, 213, 214
Hewetson, H.H. 23, 25
Hindley, Dave 99
Hobson, George 35

Hochhausen, Ben 89, 97–101
Home Oil 88, 172, 211, 230, 256
Hopkins, O.B. 9, 116, 290
Horbal, Phillip 55, 194, 200
Horsfield, Roly 87
Hosegood, Bill 194
Houston, King 167
Howe, C.D. 117
Howes, Al 52
Hoyda, Don 73
Hrynchuk, Fred 228
Hrynchuk, John 173, 228–231
Hrynchuk, Larry 228
Hrynchuk, Mike 228, 229
Hrynchuk, Peter 228, 229
Hrynchuk, Rose (nee Wispinsky) 231, 232
Hrynchuk, Sophia 32, 228, 229
Hrynchuk, Vera (nee Wosnack) 228
Hrynchuk, Walter 25, 32, 181, 189, 202, 228, 229
Hudson's Bay Oil & Gas (HBOG) 90, 165, 173
Hughes, Stan 188, 189
Hughes Rock Bit Co. 57
Hunchak, John 182, 186, 187, 193
Hunt, Don 21
Hunter, Dean 197, 198, 251
Hunter, Les 99
Hunter, Vern (Dry-Hole) 22, 79, 98, 251, 252

Imperial Leduc #1 10, 14
Imperial Leduc #2 11
Imperial Oil Ltd. 4, 9, 11, 17, 25, 36, 40, 79, 85, 88, 90, 93, 95, 97, 98, 108, 112, 118, 122, 135, 142, 143, 147–150, 153, 158, 160, 165, 170, 172, 183, 184, 196, 198, 200, 203, 207, 208, 211, 212, 218, 239, 254, 264, 267, 270, 275, 281, 283, 284, 309, 345
Ings, Bob 194
Interprovincial Pipe Line Inc. 116, 118

Jacobson, A.T. 169, 292
Jamieson, Scotty 195
Jeffery, Bernard and Kay 89, 184, 185
Jeffrey, Ken 18, 160, 199, 309

Kashower, Bob 87
Kaveler, Mr. 95
Kay, Bob 239
Keller, A.E. 95
Kelly, Bud 40, 50, 51, 89, 196
Kenyon, Alex (Sandy) 17, 19, 104, 275, 276
Kenyon, Diane 275, 276
Kenyon, Helen (now Stetsko) 19
Keplinger, Henry 95
Kerber, Maxine 193, 198
Kerber, Victor 197
Key, Anne (nee Yachimec) 181, 190, 198, 203, 233–235, 353, 369
Keyser, Herschel 52
Kidd, Fred 14, 153, 169
Killer, Fred 22, 32, 72, 254
Kimura, Chizuco and Jim 179, 236
Kiyooka, George 179, 238, 239
Klovan, J.E. (Ed) 64, 65, 373
Knebel, Mose 195
Knode, William F. 130
Knowles, R.D. (Doug) 90, 91
Krumbein, W.C. 61, 63
Kunst, Hank 29

Lakusta, Ed 112, 122, 123, 179, 201, 204, 240, 243–245
Lakusta, Helen 201, 240–243
Lakusta, Mike 240, 242
Lakusta family 179, 241, 242
Lambert, Ed 35, 246
Land Titles Act 95
Lane-Wells (Company) 85, 88, 322
Lang, John 124
Langston, J.R. 172
Laurence, Rob 42
Lavoie, Doug 18, 67, 74, 75, 255
Law, Jim 61
Layer, D.B. (Doug) 68, 141, 195, 293
Leeson, Lornie 50
Leiper, Hughie 48–52

Lewis, D.E. (Ed) 32
Lineham, Carrie and Fin 4, 9, 22, 32, 55, 57, 190, 218, 275, 276, 311
Link, T.A. 2, 71, 172, 208, 292
Littlefield, Max 63
Lonsdale, Ab 269
Lonsdale brothers 20, 144, 311
Loranger, Dr. Diane 62
Lougheed, Don 15
Loughney, Ed 98
Lowery, Major James R. (Jim) 172, 294
Lukes, Alice and Jim 185, 194, 198, 200

MacCallum, G.T. 32, 67
Mackenzie, Don 83, 136, 167
MacKenzie, Hugh 37, 38, 89, 99, 103–106, 191, 264
MacRae, H.R. (Hec) 67, 68, 195, 253, 254, 256, 257
Madill, Floyd 114
Makowsky, A. 51
Malowany, Grace 180
Malowany, Mary (nee Wacko) 180, 188, 248, 249
Malowany Virginia (Sharek) 250
Malowany Walter 189, 190, 247–250, 265
Manning, E.C. 95, 117, 157
Manyluk, Frank 97, 101–103, 125
Mason, A.D.M. (Tony) 67
Maughan, John 90
McCaskill, A.J. (Jack) and Kay 22, 24, 30, 32, 182, 198, 251, 252, 311
McCoy, Alex 63
McDaniel, Rod 89, 92
McDonald, Bert 89, 237, 247, 251, 252
McDonald, Neil Alexander 159, 160
McDonald, S. 30
McGreevy, Bill 50
McIntyre, Boyd 46, 48
McIntyre & Webster 210
McKay, John R. 87, 322
McKinnon, Fred 29, 36, 167, 169
McKinnon, Ian 164, 165
McLatchie, A.S. 68
McLaws, D.P. 136, 152, 213
McLaws, Jean 19, 67, 73, 222, 255, 266
McLennon, Jack and Lorne 147
McMahon, Frank 153, 163, 164, 174, 214, 215
McMaster, Jim 51
Meech, Hod 134, 135, 141
Mewburn, Arthur 72, 293
Miller, Archie 47
Miller, Doug 142
Mineral Taxation Act 94
Mines & Minerals Act 94
Mobil Oil 39, 40, 62
Moholitany, Metro 40
Moore, Donald 51
Moore, Dudley 51
Morin, E. 120
Moroney, Tip 30, 40, 79, 80, 83, 98
Morris, Harry Rodney (Rod) 18, 19, 21, 22, 56, 57, 67, 72, 73, 144, 147, 183, 208, 253, 255, 268, 311
Morris-MacRae, Jean (nee Robinson) 72, 73, 253, 269
Mossop, Grant 66
Motyka, Dan 112
Munroe, Leslie 131, 143
Murray, Roy 15
Mutch, L. (Jim) 51
Myroon, John 181, 190, 261, 262

Nauss, Art 33, 72, 173
Negrich, Steffie 214
Nelson, Reg 30
Nesbitt, Jack 85, 89, 194, 195, 197, 200
Nesbitt, Ruby 197
Neufeld, Hank 51
Newmarch, Chuck 142
New Superior 172
Nickle, C.O. (Carl) 151, 162, 164
Norman Wells 9, 10, 61, 111
Northern Development 32, 42
Nowers, B.L. (Lauder) 89, 120, 194, 248, 263, 269, 372
Nu-Mac 51

Oakley, Ken 89, 119, 120, 184
O'Brien, Jack 175

O'Brien, Mel 143
Oil & Gas Resources Act 94
Osborne, Tommy 184
Owre, "Slim" 20, 311

Pacific Petroleums 33–35, 88, 90–92, 163, 170, 173, 208, 214, 236, 241, 243, 270
Parker Drilling 37
Pasemco, Victor 115, 117, 118, 209, 266, 267
Patterson, Pat 53
Paulson, Maurice P. 22, 68, 81, 82, 89, 92, 106
Penn West 51, 52
Petro-Canada 112, 142, 245, 281
Petroleum & Natural Gas Conservation Board - see Energy Resources Conservation Board (ERCB)
Phillips, Glen 51
Pich, Mary 259
Pinedale (well) 29, 141
Pipelines Act 116
Placer-CEGO 46
Pooler, Stan 89
Porter, Marsh 135, 152
Pot, Robert (Bob) 85
Powell, Gene 98, 145
Pratt, Wallace 9
Proust, Emil 51
Purschke, Adolphe 183

Rabey, W.N. (Wes) 11–13, 15, 18, 29, 34, 153, 283, 301, 302
Rault, Corrine (nee Dziwenka) 224
Rebus, John 51
Redman, Don 213
Redwater Operators Committee 88, 338
Redwater Pool Producers Committee (RPPC) 94
Redwater Unit Legal Subcommittee (RULS) 93, 94
Redwater Water Disposal Company (RWDC) 96, 98, 100, 211, 281
Redwater Working Committee 90, 92, 93, 97, 338
Reid, Clifford 51
Rice, Ivan 98
Rice Engineering 99, 101
Richardson, Sandy 137
Roberts, Frank 10–12, 290
Robertson, Harry 89
Roloff, Myrt and Rudy 4, 109–111
Romanchuk, Diane and Peter 281
Ronaghan, Frank 34
Roshko, Frank 184, 201
Rosichuk, Jack 51
Royalite Oil 35, 46, 88, 90, 92, 144, 148, 165, 166, 168, 171, 173, 263, 271
Royer, Emile 17, 72, 268, 270
Royer, Eve (now Forry) 186
Royer Rita 268, 269
Royer, Yvonne 268–270
Rozak, Nick and Katherine 210, 271, 272
Rutherford, Dr. Ralph 62
Ryland, Cloyce W. ("Goat") 39

Savage, Gordon 49
Savage, Peter 152
Sawka, Mary 193, 273, 274
Schlumberger (Company) 82–85
Scott, James C. (Jim) 161
Scragg, Theresa (later Chaba) - see Chaba
Scrymgeour, John 172
Scutchings, Bill 147
Seaman, B.J. 20, 47
Senych, Mike 190, 224, 225, 227
Sexton, Whitney 116
Shatford, Reginald Alexander (Reg) 66–70, 74, 87, 89, 90, 144, 238
Shaw, Dr. E.W. (Ernie) 14, 15, 19, 66, 195, 295, 302
Shaw, John, Jr. 216
Shell Oil 35, 135, 138, 150–154, 158, 161, 169, 213, 230, 231
Shepard, W. 51
Shultis, George 80
Slatter, John 133, 137

Sluzar, R.D. (Ray) 68, 290
Smith, Bert 248, 265
Smith, Bill 89, 188, 202, 218
Snortland, Floyd 51
Socony-Vacuum - see Mobil
Somerville, H.H. (Hubert) 35, 95, 131, 157–160, 165, 169, 170, 201, 261, 350, 353, 357
Spence, Al 137
Spencer, Cody 50, 172
Sproule, J.C. 293
Starr, Ernie 104
Statler, Jim 259
Stelck, Charles R. 64
Stetsko, Helen and Matt 19, 275, 276, 281
Stevens-Guille, Henry Lamarchand 82
Stewart, G.L. (George) 117, 196
Strand, Jim 90, 91
Swann, R.H. (Rich) 88, 90

Talisman 112, 171, 230
Tanner, Harold 171
Tanner, N.E. 32, 130, 158, 164, 167, 171
Tanner, Reg 171
Taylor, Austin C. 143
Taylor, John 21
Taylor, Neil 42
Taylor, Vernon (Vern) 68, 82, 98
Taylor, Walker L. 25, 30, 32, 79, 82, 199
Tellington, Wentworth 33, 34
Teske, Gordon 114, 120
Texaco 49, 110, 139, 170
Tixier, M.P. 85, 86, 373
Tkachuk, Pete 255
Torguson, Norma (nee Anderson) 188, 189
Tosh, Scotty 34, 51, 243
Tovell, Ed 22
Tull, Louise and Ray 39, 41–43, 142
Turnbull, Hardy 76
Turner, Jack 89, 197
Twaits, Bill 83, 196

United Oils 172

Van der Grift, Millard 89, 196, 251, 252, 275, 276
Van der Lee, Ari 105, 107–109
Visser, Charlie 46, 50, 82, 83, 373

Walker, Bessie 222
Walker, Len 46, 48
Walters, Ray 10, 18, 29, 79, 81, 82, 213, 247, 292
Wanamaker, Mr. 95
Wark, Tom 53, 54
Wasylyk, John 140
Webb, J. B. (Jack) 30, 32, 81, 82, 290
Webster, Gordon 88, 91
Webster, Harry 46–48, 54, 271, 291
Weeks, Lewis, G. 9
Weller, Sid 149
Wendte, Jack 75
West Canadian Petroleums 45, 47
Western Leaseholds 135, 137, 142, 147, 227
Western Minerals 88, 90, 91, 133, 135, 137, 141, 142, 144, 147, 162
Whaley, Bill 137
White, J. R. (Jack) 116, 196
Whiteford, W.K. 165
Wilkins, Kitty (Libbey) 15, 200
Williams, C.D. 95
Willingdon (well) 29, 141
Wilson, Harry 95
Woodhead, Mike 17, 269
Woods, Rex 90
Wright, Al 293
Wright, Cyrus 174
Wright, Geo. 92
Wright, Harry 84
Wright, Marge (Malowany) 180
Wright, Norm 89

Yachimac, Andrew 202
Yachimec, Peter 179
Yarmula, Pete 190, 191, 201, 366
Young, J.W. 68, 120

Zeniuk, Nick 200, 371
Ziegler, Jim 10, 34